Praise for *Inventic*

This book is a great 'how to manual' for scientists and engineers aspiring to become entrepreneurs and change the world.

Larry first learned at Stanford that scientists can start companies, then he went into the world and did it over and over again.

Read this book and learn how you can do it as well.

> – Steve Blank, Creator of the Lean Launchpad and architect of I-Corps

Australia needs agitators. Credible, visionary leaders of billion-dollar enterprises like Larry Marshall who push us to innovate and do better. Despite being one of the top 15 economies of the world, Australia is below 70th in economic complexity, a measure of innovation. It is critical that Australia has a step-change in innovation if we are to navigate the environmental, energy, geopolitical and many other challenges in front of us. Larry is a rare breed – both a scientist and a CEO with world-class innovation credentials – and so the perfect agitator to help Australia understand how we can reinvent ourselves and our country, by showing other scientists how to walk his path.

> – Maile Carnegie, Group Executive Australia Retail, ANZ

Larry Marshall has played a game-changer innings as head of our esteemed science agency CSIRO. As a venture capitalist and startup enthusiast, his appointment was bound to be controversial. But his passionate pursuit to match up great science with great translation and commercialisation has enlivened and enhanced CSIRO's drive for heaps more innovation in Australia.

> – Bill Ferris, AC, Co-Founder of Champ and AMIL; former Chair of Innovation and Science Australia

Powered by his extensive scientific entrepreneurship, Dr Larry Marshall shows us how to couple science with innovation to produce prosperity. Human ingenuity is an inexhaustible resource; this book explains how to mine it and refine it into societal value. Listen up politicians and investors, Australia needs this work.

> – Dr Alan Finkel, former Australian Chief Scientist, President of the Australian Academy of Technology and Engineering, Chancellor of Monash University, CEO and Founder of Axon Instruments

Australia has a proud history of scientific research and industrial innovation. But despite this, we've got a lousy track record of translating this innovation into real impact, especially commercial impact.

We must, as a nation, take a cold, hard look at where we are failing, if we are to succeed.

We need to develop a culture of innovation and risk-taking. That means always having a crazy, brave Plan A, backed in by a bulletproof Plan B.

We must move away from being quick to celebrate failure, to defining success as having a go and learning from failure.

It means developing industrial-scale quantities of green alternatives to fossil fuels – green electrons and green molecules – is the most important challenge facing Australia and the world today. It also means decarbonising our vast mineral wealth – green iron and steel, lithium, copper and nickel.

We will only succeed if we try. This book is an important first step towards success.

> – Dr Andrew Forrest, AO, Chairman and Founder of Fortescue Metals Group, Fortescue Future Industries, Minderoo Foundation and Tattarang

Few scientists have transitioned to become business leaders, or to create public companies, but Larry Marshall has done just that – and by sharing uncomfortable truths, failures and successes, all anchored by the real-life experience of someone who has crossed the Valley of Death more than once, Larry seeks to provide other scientists with the confidence that, they too, can do it.

I have seen Larry disrupt, lead and inspire change, but more importantly, create and deliver new value. This book showcases how value can be created by and with science – and Australia's world-class science has certainly delivered examples of huge value-creating outcomes.

Through his book, Larry provides the incentive and encouragement for more scientists to choose this path, as indeed they must, if Australia is to benefit from the opportunities, rather than suffer the consequences, of our energy transition and decarbonisation challenges.

> – Catherine Livingstone, AO, former Chair of CSIRO, Commonwealth Bank and Telstra; former President of the Business Council of Australia; and former CEO of Cochlear

Australia has produced a remarkable number of world champions and people at the top of their game. Whether in science, business, sport, the not-for-profit sector or the arts, we have heroines and heroes in so many key places and positions. But if we are truly honest with ourselves, there is one area where we've underperformed. Whilst we've been reasonably strong in discovery, we've been weaker than we should have been in 'getting it out there'!

We're now blessed with a wonderful book from successful science entrepreneur, Dr Larry Marshall, who highlights the issue and what we can do about it.

Larry is in a unique position to have written *Invention to Innovation* because, not only was he the Chief Executive of Australia's national science agency, CSIRO, for 8 years, but preceding that he had a celebrated career in Silicon Valley including not one but two IPOs.

This is a 'must read' for anyone who is passionate about what can really lift Australia into the nation that it can and ought to be.

> – Simon McKeon, AO, FAICD, Chancellor of Monash University; 2011 Australian of the Year and former Chairman of CSIRO

Larry has worked on the frontlines of innovation all of his professional life. From many rides on the rollercoaster of startups, as an investor with the heart and empathy to support the next generation and then reshaping the whole system through his roles in government. No one is better placed to describe the ins and outs of innovation both from a historical context but most importantly from the perspective of what we need to do differently in the future.

> – Niki Scevak, Founder of Blackbird and Startmate

The Digital Future has huge potential to unlock new waves of innovation and economic prosperity for all Australians. It's a future where Aussie kids see Aussie scientists and Aussie entrepreneurs solve Australian problems and take them to the world. Larry is passionate about this future for our children, and this book is all about how to make it happen. We certainly have the talent and the track record – and we need to foster, nurture and invest in research, science and innovation at all levels of the economy to realise this potential in the decades to come.

> – Melanie Silva, Managing Director of Google Australia and New Zealand

Invention to
INNOVATION

HOW SCIENTISTS CAN DRIVE
OUR ECONOMY

DR LARRY MARSHALL

with JENNA DAROCZY

CSIRO

PUBLISHING

A catalogue record for this book is available from the National Library of Australia.

ISBN: 9781486316373 (pbk)
ISBN: 9781486316380 (epdf)
ISBN: 9781486316397 (epub)

How to cite:
Marshall L, Daroczy J (2023) *Invention to Innovation: How Scientists Can Drive Our Economy*. CSIRO Publishing, Melbourne.

Published by:

CSIRO Publishing
Private Bag 10
Clayton South VIC 3169
Australia

Telephone: +61 3 9545 8400
Email: publishing.sales@csiro.au
Website: www.publish.csiro.au
Sign up to our email alerts: publish.csiro.au/earlyalert

Front cover: illustration by maglyvi/Shutterstock.com

Edited by Joy Window (https://livinglanguage.wordpress.com)
Cover design by Cath Pirret
Typeset by Envisage Information Technology
Index by Max McMaster
Printed in Australia by McPherson's Printing Group

CSIRO Publishing publishes and distributes scientific, technical and health science books, magazines and journals from Australia to a worldwide audience and conducts these activities autonomously from the research activities of the Commonwealth Scientific and Industrial Research Organisation (CSIRO). The views expressed in this publication are those of the author(s) and do not necessarily represent those of, and should not be attributed to, the publisher or CSIRO. The copyright owner shall not be liable for technical or other errors or omissions contained herein. The reader/user accepts all risks and responsibility for losses, damages, costs and other consequences resulting directly or indirectly from using this information.

The information in this book is for general information purposes only. It should not be taken (or relied upon) as constituting financial advice from the author or CSIRO. Before making any investment decisions, you should seek independent legal, financial, taxation and other advice to check how any of the information in this book might relate to your unique circumstances. The author and CSIRO are not liable for any loss caused, whether due to negligence or otherwise arising from the use of, or reliance on, the information provided in this book.

CSIRO acknowledges the Traditional Owners of the lands that we live and work on across Australia and pays its respect to Elders past and present. CSIRO recognises that Aboriginal and Torres Strait Islander peoples have made and will continue to make extraordinary contributions to all aspects of Australian life including culture, economy and science. CSIRO is committed to reconciliation and demonstrating respect for Indigenous knowledge and science. The use of Western science in this publication should not be interpreted as diminishing the knowledge of plants, animals and environment from Indigenous ecological knowledge systems.

The paper this book is printed on is in accordance with the standards of the Forest Stewardship Council® and other controlled material. The FSC® promotes environmentally responsible, socially beneficial and economically viable management of the world's forests.

Mar23_01

Foreword

Australia is a great country in which to be a scientist. Across our universities and government agencies, we have an amazing community of outstanding discoverers – people who are curious, at the leading edge technologically, and able to do things that others can't. Yet when it comes to connecting these talents to entrepreneurship, we hold ourselves back. We are much too cautious in striving for commercial impacts. There's a weird cultural belief that great companies and great research cannot be led from these shores.

This book is an antidote to such views. For Australia's budding technology entrepreneurs, it is an excellent 'how to' manual, full of practical advice, and offering useful, tangible guidance on how our scientists and entrepreneurs can seize the tremendous opportunities Australia offers. Much of this is based on Larry's own, first-hand experience.

Along the way, he synthesises a series of broader insights, including one that resonates strongly for me: the idea that Australia should stop replicating the innovation systems of other nations and leverage our own strengths. We really do have a unique and advantageous culture in this country: we do not shy away from taking on the hardest challenges; we have an incredible record of adapting technologies for our purposes; and we are culturally extremely effective at growing collaborative teams to devise imaginative solutions to our problems.

These are precisely the traits that one would hope for in an innovative society. Larry's unswerving passion and conviction in Australia's ability to become globally competitive in innovation is inspiring. Perhaps, as others read this book and word gets out, more Australian scientists will come to realise how they can convert their inventions to innovation. I wish that I'd been able to read it years ago.

Michelle Simmons, CEO and Founder of Silicon Quantum Computing and 2018 Australian of the Year

Contents

About the authors

Dr **Larry Marshall** was Chief Executive of Australia's national science agency, CSIRO, from 2015 to 2023, and prior to this spent more than 25 years in Silicon Valley, founding, leading and investing venture capital in deep-tech companies. With a PhD in physics, 20 patents, six startups and two IPOs to his name, Larry is a passionate supporter of Australian innovation and the power of science and technology to drive our economy and build resilience to future challenges.

Jenna Daroczy believes that well-chosen words and stories can inspire change. She works with leaders and changemakers to articulate their vision in robust ways that ignite interest, build understanding, influence opinion and inspire action. Jenna has worked as a journalist and in corporate communications for nearly two decades and has qualifications in communications, science, law and policy. She is currently the Leadership and Strategy Communications Manager at Australia's national science agency, CSIRO.

Introduction

In the early 1990s, researchers in radio physics at Australia's national science agency, the Commonwealth Scientific and Industrial Research Organisation (CSIRO), were working on solving a problem in radio astronomy. They were trying to find a way to stop signals reverberating off objects and creating interference which, in this instance, was making it hard for them to study black holes. When they eventually found a solution, they realised it could also be applied to other similar problems, like those facing the makers of the increasingly popular portable computers at the time, who were trying to find a way to connect to the internet without having to be plugged in with a cable.[1] In 1996, the team led by Dr John O'Sullivan, and including Dr Terry Percival, Diet Ostry, Graham Daniels and John Deane,[2] secured a patent for Wireless Local Area Networks (WLAN),[3] which would go on to be one of the technologies enabling what we now call wi-fi.

There are a few 'urban myths' about how CSIRO came to invent fast wi-fi (wi-fi that can operate at a required speed to be useful). The one I hear most often is that wi-fi wouldn't have been discovered if the scientists hadn't been allowed to do whatever research they felt like – but that's not what happened here. While the team didn't set out to invent wi-fi, they *were* trying to solve a problem. CSIRO's lead on the wi-fi litigation, Dr Jack Steele, recalls:

> The Division was responding to the then government's desire for CSIRO to increase its external revenue. The Division did a strategy process to identify new technologies that, if developed, would have high commercial value and so high IP [intellectual property] revenues for the Division. Wi-fi was selected as one candidate to work up (circa 1991/2). Would the astronomy research have led to this outcome if there had not been that intervention and focus on commercial outcomes? Questionable, I would have thought.[4]

The fact that it took asking questions about black holes to find answers to connectivity issues in our homes just speaks to the incredible power of science to deliver solutions in unexpected places – but you do have to be asking questions in order to find answers. The research team who patented the WLAN technology had been working with Dr Dave Skellern and Dr Neil Weste at Macquarie University, who went on to develop a microchip that could be used in computers to connect wirelessly and commercialised it by founding the company Radiata Communications in 1997.[5] Reflecting on that time, Dave says that characterising the use of the chip as being for 'computers' is only part of the story:

> It doesn't even hint at the breadth of the vision we were aiming to enable. Our chip specs were designed for wireless internet connections to anything, in home, public spaces and workplaces. We had development contracts with Cisco for enterprise, Broadcom for set-top boxes and other home networking products, Sharp for entertainment products and Skypilot for community and public hotspot equipment. We had reached agreement but not yet signed an agreement with Nokia for use in mobile phones.[6]

Trying to raise investment for the company in Australia proved futile, but as Dave recalls, Australian venture firms' 'terms were lousy and they didn't give us any connectivity, they didn't give us any relationships that we didn't already have in the US and worldwide. We knew close to all, if not everyone, in the game – our competitors and our potential customers.'[7] However, they had no trouble raising funds in Silicon Valley where Dave says they could work with 'trade investors who invested in our company as well as giving us contracts for development of products that we knew would actually get into use'.[8] By 2001, Radiata had been bought by US technology company Cisco for $565 million.[9]

As CSIRO retained the patent for the breakthrough, it began trying to encourage industry to take out licences for its patented technology. Perhaps unsurprisingly given the warp speed of internet adoption, CSIRO did not succeed in having its technology licensed as a proliferation of companies seized whatever technology they could to connect to the internet. Many likely thought there would never be enough resources for all of them to be pursued, and perhaps underestimated the interests of an Australian research organisation to do so. In 2005, CSIRO began what would become many years of court battles in the US to claim royalties from technology companies using its patented wi-fi technologies, which have since reaped hundreds of millions of dollars for Australian science.[10] In 2009, my predecessor as CSIRO's Chief Executive, Dr Megan Clark, delivered $150 million from WLAN royalties to the Science and Industry Endowment Fund[11] (of which I am currently Trustee now as CSIRO's Chief Executive), which funds Australian research projects delivered with industry partners. After I took up the role as Chief Executive in late 2014, CSIRO negotiated the last $100 million of the WLAN settlement out of court with the remaining litigants, which helped establish the CSIRO Innovation Fund in 2015,[12] discussed further in Chapter 6.

Fast wi-fi was an amazing breakthrough that generated massive value and is a gift that has kept on giving. It is one of CSIRO's – and Australia's – proudest inventions, enabled by a team of five brilliant researchers whose work with complex mathematics and understanding of radio waves unlocked a truly game-changing solution to a rapidly growing challenge. But if you ask most people today who invented fast wi-fi, they won't be able to tell you it was an Australian team, let alone CSIRO – in fact I shocked international innovation expert Professor Mariana Mazzucato when I mentioned it to her a few years ago. I think the reason too many people don't know about the origins of wi-fi is because CSIRO didn't succeed in commercialising the WLAN patent here in Australia all those years ago. CSIRO didn't create a product we could sell to companies that came with a 'Made in Australia' sticker on it. We didn't start a company that grew jobs and economic wealth for Australia using that technology and could go on to become a household name like Telstra or Qantas. I don't blame anyone involved – after all, it's still incredibly hard to commercialise research in Australia today, nearly 30 years later.

Fast wi-fi should be rightfully celebrated as a great Australian *invention* that paved the way for a paradigm shift in technology. But it should not be celebrated as an Australian *innovation* – an innovation would have taken the invention past the point of discovery through to delivering economic benefits for Australians. Wi-fi should have enabled the formation of a series of great Australian companies, one creating the magical semiconductor chip like Radiata, another building the wireless modem like Netgear, and another building the telecommunications interface like Belkin. Any one of those would have been great for Australia by realising their value here, creating an industry here and employing Australian graduates here. Innovation is all about learning from failure, but Australia has been slow to learn.

So it's not surprising to realise most of our innovators learned the lesson overseas. In the final year of my physics PhD, I had the opportunity to go to Stanford University in California to complete my research. It was the late 1980s and Silicon Valley was a thriving hub of investment in the latest software and early computer technology. As a student on campus, I met many venture capitalists – investors in high risk, early stage companies – who would spend time in Stanford's labs, learning about new technology and looking for their next investment. It was this close relationship between scientists, entrepreneurs and investors that drove the Valley's world-leading innovation in supporting science across what is called the 'Valley of Death': the challenges that lie between a brilliant breakthrough in a lab and the market-changing product arriving in a customer's hands. The 'Valley of Death' is a phrase used to describe the funding shortfall a startup faces between investing in developing its invention into a product and being able to start generating revenue from getting the product into market.

In Australia, we actually have three Valleys of Death, respectively caused by our culture, our capital, and our customers, as shown in the following figures. The first Valley of Death reflects our cultural focus on strong academic performance, which has seen Australia continually deliver world-class science excellence (Figure 1). We love great science, which is wonderful – but high academic excellence, measured here by NCI (normalised citation impact), is usually low on commercial readiness, measured here by TRL (technology readiness level). To commercialise research, you need high TRL.

The second Valley of Death reflects our national preference to put our capital in low-risk, high TRL investments, like infrastructure assets (Figure 2, next page). That means less capital available to go into commercialising science. Putting that into context, in 2017 Innovation and Science Australia (ISA)

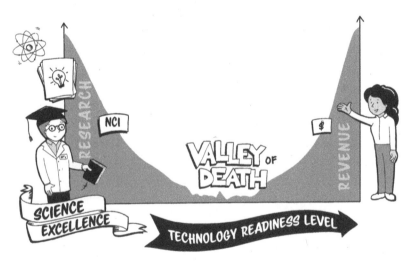

Figure 1: The first Valley of Death - invention versus innovation. Illustration by Debbie Wood.

reported that Australia spends about $10 billion each year on research. But our $2.5 trillion capital market[56] won't invest to support that science across the Valley of Death – it will generally only invest once the science has been successfully commercialised.

The third Valley of Death is the most common one that all ecosystems have and reflects different customer bases that drive a company's growth – the gap between early adopters and the mass market that brings in sustainable revenue (Figure 3, next page).

Digital startups can fly over many of these challenges because they iterate so quickly from direct customer feedback without having to go and physically meet a customer or deliver the product, but the Valleys are deep and wide for deep-tech startups. All three of these Valleys compound to create our 'innovation dilemma': Australia's world-class research doesn't result in commercialisation-ready inventions; those inventions are often too high-risk for our existing investment community to support; and even when we do support them, we don't usually

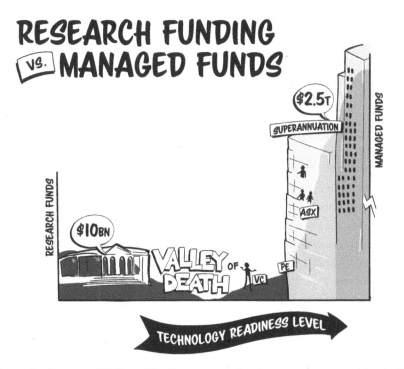

Figure 2: The second Valley of Death - research funding versus managed funds. 'VC' is venture capital; 'PE' is private equity; ASX is the Australian Securities Exchange. Illustration by Debbie Wood.

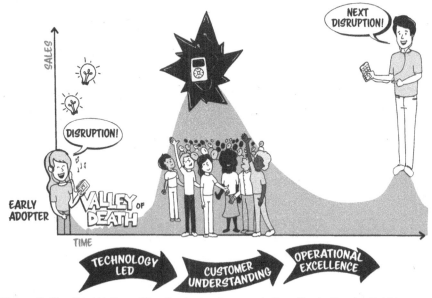

Figure 3: The third Valley of Death - technology evolution. Illustration by Debbie Wood.

support them all the way through to sustainable revenue, so either they fail or someone else invests and reaps the value. The simple fact is this: if you just invent but don't build the company, your value will be diluted away to nearly nothing by the time the company goes public and real value is created. To capture real value from commercialisation, we must take the invention all the way to product and take the product all the way to market – we must invest in the company and stay deeply engaged as it commercialises in order to retain value. Creating startups is much harder than other commercialisation options, like simply licensing the technology, both for investors and for IP-generators like CSIRO and universities. Venture capitalists don't want to have to pay royalties to research organisations for the life of the company or to have them on the startup's capitalisation table – they want to invest all their money into the people focused on making the company successful *after* the invention, not before, so equity is a tough model for universities to recoup their investment into the initial research. Royalties might do that, but they don't create jobs or new industries for Australia the way startups can.

Because Australia has three Valleys of Death, crossing them doesn't just take a bit of investment; it also needs a strong, diverse and complex innovation ecosystem that can shepherd science – and scientists – to the other side. It needs a culture where science can pursue big, bold, 'blue sky' research as well as targeted solutions to real-world problems, because you can't have one without the other, as wi-fi demonstrated through its origins in radio astronomy. That culture creates an appetite in industry to invest in innovation and see it as critical to business strategy, not just as a 'nice to have' if there's spare cash. A strong innovation ecosystem is broader than just research and industry – it needs leadership that invests in big ideas and sets clear expectations about its direction, as well as an investment community, often in venture capital (VC), with a healthy appetite for risk, based on an understanding of how science and technology operate. All of these things working together create an environment that inspires the next generation of scientists and creates opportunities for them to pursue challenging, exciting tech careers.

Australia does not currently have a thriving innovation ecosystem that can launch its ideas across the Valley of Death and into commercial reality, creating jobs, growing markets, and strengthening our economy through solutions and products that make life better. Australia's Valley of Death is one of the main reasons wi-fi didn't become a commercialised product, let alone creating a thriving Australian company. Instead, we have an innovation dilemma where our scientific research is world-class – Australia is in the global top 10 for science[13] – but we lose our best ideas to be commercialised overseas where thriving innovation ecosystems are powered by VC, government investment, industry engagement and world-class research, which all come together to form bridges across the Valley of Death. When we lose those ideas to be commercialised overseas, we lose economic, social and environmental benefits with them.

But I believe Australia can have a system like that, and it's not as far out of reach as it might sound when you consider the wi-fi story. This book is full of stories like wi-fi, of Australian ideas lost to the Valley of Death, as well as how some companies successfully crossed it. I've spent my career living the Australian innovation dilemma and trying different ways to solve it. This book builds on my experiences from my time at Stanford, to being a founder of science-driven startups in the US, to investing in American, Australian and

Chinese science-driven startups as a venture capitalist and, most recently, my time as the Chief Executive of CSIRO. It looks through the lens of these experiences to identify what we need to do to build a uniquely Australian innovation system that allows our research to leap over the Valley of Death with confidence.

We won't be successful – or innovative – if we try to copy what other countries have done, but we can learn from the insights, successes and mistakes of other nations, like Silicon Valley in the US, to build a uniquely Australian innovation system that reflects us at our best and solves the problems that matter most to us. We have our own potential, our own strengths and our own opportunities, and can create our own unique ecosystem. We might be a small population, but we are a smart one. While we cannot compete on size or scale, we can use science to play to our advantages by reinventing those strengths to solve the seemingly impossible, because that's what Australian science has done, time and time again.

Part 1: How to transform a scientist

What stands between a scientist with a brilliant idea and delivering real-world impact from that idea? What prevents commercialisation of an idea that will solve a huge problem, create jobs and economic benefit, or even create a dramatic paradigm shift with a new breakthrough technology? The Valley of Death. To leap across the Valley of Death, you need financial support (often in the shape of VC), leadership (not necessarily in the traditional sense of business leaders) and a strong 'market vision' of how this idea will fit into and improve the lives of your customers.

Countries with a reputation for strong innovation, like the US, Israel and China, have created ecosystems that enable scientists with great ideas to cross the Valley of Death. In Australia, our fledgling innovation system means many scientists leave the country for a more innovative ecosystem, or they're simply unaware that attempting to leap across the Valley is even something they could, or should, try to do. I was one of those scientists who left Australia to turn my ideas into real-world products, and in the process I founded six companies and became one of the venture capitalists I'd relied on to get my start.

Part 1 of this book draws on my experiences as a scientist working in, founding and investing in deep-tech startups from around 1988–2014 to suggest how we can help our scientists to leap across the Valley of Death at home in Australia. It looks at why our Valley of Death is currently so hard to cross; how our growing VC industry can best support scientists with deep-tech ideas; how our scientists can become a new kind of Australian CEO; and what it takes to develop a powerful 'market vision' that will help those companies – and the venture firms that invest in them – to succeed. At the end of Part 1, I've also included some starting points for scientists thinking about raising VC.

The power of science to solve our greatest challenges, grow our economy and reinvent our future is held in the hands of our nation's greatest untapped asset: our scientists.

1

Australia's innovation dilemma

My story

I've always been passionate about using science to solve problems. When I interned at CSIRO in 1984 at the end of my physics undergraduate degree, my supervisor John McAllan taught me, 'If you don't deliver it, you haven't really done it.' He meant that you can claim your science can solve a problem, but until you've implemented the solution, the problem isn't fixed. CSIRO had been created at the turn of the century to solve the nation's prickliest problems – including the literally prickly problem of prickly pear in the 1920s[14] – in partnership with industry, universities and government. While I had always understood science to be a tool that could be used to solve someone else's problems, I had no idea it could be an industry in and of itself. I don't think many people in Australia really think of science that way either, or at least think that it can play that role in Australia's economy.

My PhD supervisor at Macquarie University in Sydney, Professor Jim Piper, set me on the course to discover that science can be more than just a tool to solve problems – it can be an entire pillar of a nation's economy. Throughout my studies, Jim inspired me to seek the purpose of science and ensure what I did became relevant in the everyday world. He took his students to pre-eminent overseas conferences where he convinced us we were every bit as good as the great laser researchers at places like Stanford – it was transformational for young PhD students to be inspired in this way. He taught me that in science there is no failure, but simply learning and trying again. His underlying philosophy was, 'It's not only the science that matters, but what you can do with it.' Jim leveraged his former students, who had established themselves around the world, and sent more of his students to visit them, exposing them to different ways of thinking and the global research and development (R&D) perspective. Many of his students became connected with Oxford, Stanford, Cambridge and Princeton, among others, through him. In my case, he connected me with Professor Bob Byer, Dean of Research at Stanford, and convinced him to be my PhD examiner. This is how I came to move to the US in 1989 for the final year of my studies.

At that time, Stanford was synonymous with the booming innovation hub of Silicon Valley. Intel's Mountain View offices, which had opened 20 years earlier, had created a thriving ecosystem of science-driven startups. These growing businesses were rewriting the books universities were teaching in business classes around the world, while students in Stanford's business classes were proof-reading the next generation of lessons. Eager to soak up every opportunity this new experience afforded, I sat in those Stanford business classes and learned not just about venture, but also about the role of science and deep-tech in transforming the very Valley I was living in. I became hooked on the idea of using my science to be part of this dynamic world, instead of being confined to the lab bench and handing over ideas to others to apply as they saw fit.

I'd met people who worked in the growing field of VC because they were often visiting researchers in the labs. They had a thirst for knowledge about the latest breakthroughs and a keen instinct for where they could be applied to a product or industry to disrupt, and often grow, the market. They had some interesting ideas about my lasers, but my work wasn't really ready for commercialisation at that point. What was more interesting to me at that stage was the career path I saw some of my fellow scientists taking into tiny, fledgling companies of only a few people: a startup. They were part of the startup because of their technical skills and their passion for their invention, but it seemed like everyone in the startup – regardless of their background or expertise – had a hand in building and shaping the company. In Australia, I'd never heard of someone contributing to leading a business without a formal business background, but I was keen to try.

When I finished my PhD at the end of 1989, I had to choose between getting on the university teaching track towards tenure – either by returning to Australia or taking up one of the offers I'd received from Oxford and Stanford – or staying in the US and trying my hand at some of the business lessons I'd soaked up. I knew returning to Australia would not give me the same commercial opportunities as the US. I could try to apply my research somewhere like CSIRO or the Defence Science and Technology Organisation – given I'd had stints working at both – or I could apply for one of the few industrial research jobs in the private sector. I figured why not give the US a year and see what I could achieve? Over the next two decades, I co-founded six companies to commercialise research: Light Solutions, Iridex, AOC Technologies, Lightbit, Translucent, and Arasor all discussed in this book.

I'd spent the first few years on the east coast in Reston, Virginia, watching web portal provider AOL (then called America Online) be born, but soon gravitated back to the Valley where the venture community had an insatiable appetite for 'deep-tech' – technology invented by science, and where the mentoring ethos of VC helped me to rapidly increase my experience as a rookie CEO.

In the early 2000s, I came back to Australia to meet with state and federal governments to discuss innovation as part of a tour group of expats living and working in Silicon Valley. That was when I really started to understand the depth of the innovation problem for Australia. In the 10 years since I'd graduated and soaked up the lessons of Silicon Valley, not much had changed for scientists with innovative ideas back home. We certainly weren't teaching them that they could start companies to commercialise their creations. That ignited my passion to do something to change the situation and led to a few of us founding Southern Cross Venture Partners and then Blackbird Ventures, both discussed further in Chapter 2.

After 26 years in the US, the opportunity arose to return to Australia to lead CSIRO. One of my partners jokingly emailed me the advertisement for the Chief Executive position. While it wasn't intended as a serious suggestion, it got me thinking that maybe it could be the innovation catalyst I'd been looking for to solve Australia's innovation dilemma. It may seem unusual, but I wanted to solve Australia's innovation dilemma so much that I gave up great VC funds like Blackbird and Southern Cross and packed up my family in California to move them around the world to Australia. Though my kids may not have appreciated it at the time, I was doing it for them – and all our children's futures. Climate change is regarded by many as the greatest threat to humankind, but I believe Australia won't fix climate change without first solving our innovation dilemma. I also believe many of our smartest children will leave Australia to get great innovation jobs overseas. Neither of these realities creates much of a future for Australia. As Australia transitions to a net zero future, reducing our income from mining and fossil fuels, we can either lose a third of our economy or turn the disruption into an opportunity to create a new economic pillar from innovation, driven by our excellent science and ability to solve our greatest challenges, including climate change, to give all our children a better future right here in Australia.

I believe CSIRO is essential to fundamentally changing how innovation is done in Australia, but it has been a culture shock for me on many levels. I had moved from heading up small startups to now leading more than 5500 people, engaging with politicians and the public service – but it also felt like coming home. That's not just because I was thrilled to be back in Australia and raising my teenagers here, but because people at CSIRO are driven to use science to make the world a better place and that has always been at the heart of my research and innovation. But that's not to say I didn't disagree with people along the way. I was the Silicon Valley upstart who dismissed 30 years of innovation reviews, led by some of Australia's most influential people, as not providing the right answers. I was the entrepreneur (a dirty word when everyone else in the room is a scientist) suggesting that our national science council wasn't asking the right questions. Worse still was the perception of being the 'wrong kind of VC' – a venture capitalist, not vice-chancellor – who was going to 'shark tank' and pillage a national treasure. And I was the name in the media taking responsibility for decisions that weren't always popular, and sometimes weren't even mine, because I believed in doing things differently.

I must have stepped on every landmine and tripped every tripwire there was – not to be deliberately annoying, but without knowing what I didn't know. I didn't know the history and I didn't grow up in the ecosystem. But eventually people realised that I wasn't saying and doing these things to criticise anyone; it was because I was coming from a different experience – and might have some answers. In particular, a number of departmental secretaries realised my positive motivations and took me under their wings a bit, teaching me about how to interact with stakeholders in a constructive way. Thinking differently is vital to any ecosystem, but particularly an innovation ecosystem, no matter how uncouth or arrogant you might think the person who is providing that input.

I can see how it was difficult initially for them to peg me: was I scientist, entrepreneur or venture capitalist? I couldn't possibly be all of them at once, could I? Over my career, I've lived the solutions to every step of Australia's innovation dilemma: from being a scientist who commercialised his research overseas to avoid Australia's Valley of Death, a venture capitalist who shepherded many others through, to finally being 'inside the tent' at the government agency that I believed would be pivotal to solving this dilemma – though CSIRO could not do it alone. Part of the reason I still believe we can solve this dilemma is because history shows us that innovation is in our DNA.

Our innovation story so far

Australians have historically been phenomenal innovators, going back at least 65 000 years: from rendering poisonous seeds edible to the aerodynamic genius of the boomerang to the environmentally attuned practice of 'firestick farming', which still informs CSIRO's controlled burning practices today. Australia's first peoples invented incredible breakthroughs to support life down under.

Even as Australia became more connected to other nations, we took pride in our own ingenuity. A hundred years ago, our crops and livestock were being wiped out by the scourge of the invasive prickly pear, wool couldn't be woven, pleated or even washed for clothing, and cotton wouldn't grow in our soil. Through the efforts of Australian scientists working with farmers and governments, Australia identified and introduced an Argentinian moth to devour the prickly pear, we invented new ways of working with wool to create wool-friendly washes and tailor-friendly permanent pleats, and we have spent a century improving cotton crops that grow in increasingly dry and pest-plagued conditions, while still delivering a textile that is consistently quality-rated in the top 20 of the world. We've solved problems as broad as this wide, brown land itself: we've controlled pests like rabbits with myxomatosis and tackled flies with dung beetles; we've reinvented industries like cotton and barley to give Australia an unfair advantage; and we've transformed the world with breakthrough inventions like ultrasound imaging. Australia has even created a few innovation-led companies, like ResMed and Cochlear in the 1980s and Atlassian and Canva more recently.

But as we've become more connected to the rest of the world, our reliance on Aussie ingenuity has waned. Instead of taking pride in our powers of innovation, we're now proud to be early adopters of solutions from across the seas. In 2017, I gave a speech at the National Press Club in Canberra and referred to a then-recent Australian Innovation System report, which I think highlighted our fundamental reticence to be truly innovative by including numerous definitions of innovation. One definition meant 'adopting someone else's innovation'. This is the 'new to business' type of innovation, which refers to businesses that implement innovations they've seen others in their industry use. Nineteen per cent of Australian businesses copy innovation. In contrast to this, 'new to market' innovation is when a business invests in its own novel products. Only 5.5 per cent of Australian companies do this,[15] meaning we are no longer innovation leaders, but instead innovation followers. I have a narrower definition of innovation that refers to the new value created by deep-tech, discussed in Chapter 3 where I talk about the relationship between hardware, software and digital technologies. Former Cochlear CEO and former CSIRO Chair Catherine Livingstone argues Australia should be comfortable with more flexible definitions of innovation that reflect our strengths by being a 'technology integrator':

It was an insight I gained when I did my Eisenhower Fellowship in 1999 when the whole dot com theme was booming. I had a very interesting session with a couple of people at Rand Corporation. They drew the distinction, in terms of economies and countries, between technology makers, technology integrators, and technology takers. The US as a general proposition is a technology maker. Australia's then-Chief Scientist Dr Robin Batterham was saying that we should be effectively a technology taker, do the fast following, don't worry about developing it ourselves. And yet, my view was that the technology integrator zone was Australia's sweet spot.

Cochlear was certainly a case in point of that, taking a large number of different technologies, whether it was the software or the hardware, and integrating them to develop the cochlear implant, which was an incredible innovation, not just because of the technology, but the marketing and all of the other dimensions. But it certainly wasn't a technology taker, as in, just copying what someone else had done or fast follower. It was really putting some smarts into the way we put it together. Australia as a culture has been good at problem solving, both pragmatic problem solving, but also sophisticated problem solving.[16]

The first time I heard Catherine say that I reacted badly, but then I looked around and realised it is actually an advantage we are already turning into reality. I talked to some of the big Silicon Valley companies and found Symantec, Cisco and even Google have learned that Australia is a great test bed because our companies are hungry for their technology and are willing to take a chance on it. We're a Western market, so there are similarities in consumer behaviour and regulatory settings that can provide valuable insights to the big tech companies, but if the tech fails here, it's not as catastrophic as if it fails in a larger market like the US. Catherine's point stands up to scrutiny, but I think we've got to dig a bit deeper and try a bit harder to find out where we can be more than an integrator and instead lead the world.

Managing Director of Microsoft Australia and New Zealand Steven Worrall points to 2022 research from the Productivity Commission to build on the point:

Australia is an early adopter of technology and I've certainly seen that in my career. But if you compare us to our international peers, it's stark how we start at a reasonably strong level in terms of adoption and use of broad-based, digital cloud-based technologies but then drop very rapidly down the table when it comes to more advanced technologies – such as some of the advanced AI [artificial intelligence] and other data services.

I think that is aligned with the reality of the composition of our economy. Our dependence on mining and financial services – and an absence of deep manufacturing or any other advanced industries –

means we're less likely to innovate with advanced technologies in large sectors of our economy compared to other nations.[17]

Before the impacts of the pandemic made the numbers hard to compare year-on-year because of the scale of disruption, businesses were continuing to spend less on innovation.[18] Governments were spending a little bit more, but instead of investing in solving specific challenges with innovation, they were investing in tax rebates and programs like the Cooperative Research Centres (CRCs) to leave it to industry to decide where to invest – which takes us back to the problem that businesses are investing less.

We in Australia can compare ourselves to other Organisation for Economic Co-operation and Development (OECD) nations by considering these investment levels as a ratio to gross domestic product (GDP) (noting that nations' GDP also fluctuates from year to year, so this is not a perfect science[19]). As a nation, our total investment in R&D (measured by GERD – Gross Expenditure on Research and Development) declined from 2.11 per cent of our GDP in 2011–12 to 1.79 per cent in 2019–20,[20] well below the OECD average of 2.52 per cent.[21] Business R&D has fallen from 1.2 per cent of GDP in 2011–12 to 0.91 per cent in 2019–20.[22] Compared to OECD peers, Australia has also captured a third less value from digital innovation by not applying the latest technologies to real-world challenges to realise their value.[23] Only 1.6 per cent of innovating businesses in Australia collaborate with our world-class university research[24] and just 2 per cent of research journal articles were published with academic and industry co-authors in 2017 (ranked 25th out of 35 OECD countries).[25]

The World Intellectual Property Organisation (WIPO) publishes an annual global innovation index comparing 141 countries across a range of metrics. When I started at CSIRO in 2015, Australia ranked 17th in the index, including ranking 10th for our R&D, which captured inputs like gross expenditure on R&D as a percentage of GDP, university rankings, and number of researchers in the country, but ranked 99th for 'knowledge diffusion', which captures outputs like intellectual property (IP) receipts and high-tech net exports.[26] By 2022, Australia had slid seven places to rank 25th on the innovation index, including sliding two places to 12th on R&D inputs, and although we lifted nearly 30 places in knowledge diffusion outputs to 72nd, that's still in the bottom half of the world. Our research excellence and other inputs are strong, but our knowledge diffusion and technology commercialisation outputs are pulling us down. As I'll discuss later in this chapter – and this book – it's outputs that matter.

Over the next 5–10 years, 50 per cent of employers expect an increased demand for science, technology, engineering and mathematics (STEM)-trained professionals,[27] but education trends in STEM indicate that even if we build a strong innovation economy, our future leaders will not have the skills to rebuild back to the levels of innovation we used to have nor to participate in tomorrow's workforce. In Australian schools, enrolments in STEM subjects are at the lowest levels in 20 years with the number of school students studying STEM in later secondary (years 11 and 12) plateauing at around 10 per cent or less.[28] The long-term trends also indicate students' performance in STEM subjects is slipping.[29] Our talent pool is limited by gender inequity in STEM education and careers.[30] STEM and digital skills are essential to realising Australia's innovation and productivity potential. STEM education complements the development of critical thinking, creativity, collaboration and problem-solving, which will all be needed in future careers when more manual tasks are automated and more human skills are required. It is vital that Australia keeps pace with the technological change to advance the economy and invests effectively to develop students' STEM and digital skills.

While our OECD peers have focused on innovation and the shift to a knowledge economy, we have languished last. It may not feel this way because our economy has not known a recession in almost three decades, which is also likely to be the cause of our innovation reticence. The last 30 years of economic growth have been driven by digging our wealth out of the ground. Mining's revenues have created a strong corporate tax base but have lulled us into a false sense of security, leading to falling business investment in R&D and plateauing government investment. Our recent history shows Australia has developed a habit of taking the easy road: we dig big holes in the ground and sell what we dig up as a commodity overseas, but then buy it back as a finished product at 10 to 100 times more cost. Why couldn't we manufacture products like batteries and electric or hydrogen-powered cars here instead, which would create jobs and wealth while speeding our transition to a low emissions economy?

If we had invested in innovation, we could have created a stronger, more sustainable economic pillar from renewables than one built from finite resources. Our record economic growth has made us innovation-complacent and sceptical about the value of innovation. Or worse, it's made us fear it. Innovation has become synonymous with automation – which in turn has become synonymous with up to 40 per cent of jobs being lost – and not just for us, but also for our children who aren't learning the skills they'll need to keep up. Australians don't look to science for solutions the way we used to anymore. Even if they knew the Valley of Death was stopping our great ideas from being turned into solutions at home, would they care?

Our innovation barriers

It may have been easy to slip into non-innovative ways as we rode on the back of the resources boom, but we will need to overcome a number of challenges to return to an innovation-led economy. Before we can hope to improve, we need to be clear on what we're measuring, and I think Australia is not capturing the full picture when it comes to innovation. Part of what's holding us back from clear-cut metrics is our fear of failure. This not only makes us cautious in measurement, but it also makes us quick to criticise failure as a nation, heavy-handed when it comes to corporate governance, and risk-averse in our venture ecosystem, all of which stifle innovation. When we try to break new ground by encouraging entrepreneurial behaviours or innovative companies, we encounter cultural cringe, or an aversion to ideas like innovation and entrepreneurship, that dampens momentum for change. It's hard to tackle that cultural cringe when we don't have a clear 'market vision' to inspire and direct our science, industry and policy priorities. And finally, even if we overcame all of those challenges to inspire a new culture of innovation, we are still deeply competitive with each other, when it will take collaboration and cooperation for a nation of our size to make a mark on the global stage.

The wrong metrics

Earlier in the chapter, I mentioned a range of metrics to show how Australia is performing in innovation. The metrics we measure in Australia capture inputs like investment from government and industry into research, participation in the Research and Development Tax Incentive (RDTI), and 'research efficiency', which measures journal publication rates against investment in research.[31] By comparison, metrics tracked by international organisations, like WIPO and Knowledge Commercialisation Australasia (KCA), look at outputs that we might more strongly correlate with a definition of innovation. For example, in 2019 KCA found Australian research organisations generated over $176 million in commercialisation revenue, and 42 new spin-out companies and startups were created.[32]

Many of Australia's innovation interventions have been aimed at getting universities to collaborate more with industry and commercialise more research, but most universities do not commercialise as part of their core business. For example, data from Knowledge Commercialisation Australia in 2021 showed that the percentage of income that Australia's Group of Eight universities generated from commercialisation activities, like licensing and equity holding, averaged only 7.3 per cent. This calculation doesn't include student fees, which would see the percentage drop further.[33] Even at Stanford, which is held up as a world-leading university for technology transfer, revenue from technology transfer as a percentage of total revenue (including student fees) during the 2020 Australian financial year was less than 2 per cent.[34] Commercialisation is not part of the university's core business – in fact, it can act against the core business by shifting focus from low-TRL, high NCI research to high-TRL, low NCI research (our first Valley of Death, see Figure 1, page x). Attempting to change universities' business model is the tail wagging the dog. Interventions like R&D tax credit and CRCs have been great to increase industry funding of university research – but the problem we are trying to solve isn't a lack of funding for universities. We are trying to create more highly skilled, high-paid jobs for our children by creating new value from that great university research – not simply fund more research – and that requires a strong ecosystem around a university. Stanford has one of the world's best reputations for commercialisation because it is located in the heart of Silicon Valley – it would not perform as well if it was transplanted into a different geography.

Measuring inputs is useful to understand interactions in the system – but they are also too safe to be meaningful or instructive. There's security in tracking the same metrics for innovation year in, year out because you understand them well enough to know roughly what they're going to do under any given intervention so you can operate on a 'no surprises' basis. Part of the reason I was dismissive of 30 years' worth of reviews of our innovation system was because they consistently found a lack of collaboration between industry and research as the problem, so we consistently measured this collaboration through the same metrics: money spent by industry on R&D activities, number of participants involved in CRCs and proportion of R&D income in research organisations, such as universities and CSIRO, that come from non-government sources. Consistently these showed very little change. Basically, we measure inputs – money spent and meetings held.

Emeritus Professor Roy Green, Special Innovation Advisor at the University of Technology, Sydney acknowledges there may be flaws with measuring 'research efficiency', but notes all countries are measured in the same way, 'so if there are problems or flaws with the rankings, then it applies to everyone'. More broadly, though, he would like to see an additional input measured to contribute to our innovation performance:

> The rankings miss something in Australia, especially measuring collaboration, where we are right at the bottom of the OECD table, by not measuring consulting arrangements. The metrics capture very formal university business collaboration, but a lot of academics do consulting with companies and that isn't counted. It's a huge area of collaboration and it's often favoured by companies because they don't want to go through big bureaucracies in the university, they single out a few interesting people and say, will you work with us? Often they do, and that's something which ought to be seen as a collaboration, but it isn't counted by rankings.[35]

Catherine Livingstone also takes issue with the nuances of measuring inputs, saying:

> Collaboration isn't a helpful descriptor of what we need, it's not active enough and it's 'all care and no responsibility'. Engagement is probably a better word, because you're not going to get engagement unless there's purpose behind it. Engagement says, you really must have something of value to talk about, whereas collaboration says we can all come and work together and hopefully something will happen. It's too nice.[36]

Coming back to Australia from Silicon Valley, I'd seen and, importantly, lived a different system of innovation, whereas past reviews of Australia's innovation system were largely written by people who were and are absolutely brilliant, but they were writing about something someone else had done, not something they'd done themselves. If you're going to drive a change, you need to have lived the change personally, otherwise you can't really be authentic about it. Interventions on either side of the Valley of Death – like CRCs to make research a little more commercial, or the R&D tax incentive to make industry a little more interested in research – are just fiddling on the edges because they don't move us towards measuring meaningful things.

Because we measure inputs, we incentivise inputs. Universities are rated and funded based on journal papers published and student enrolments. Of course, researchers would love to see their work go on to make the world a better place, but first and foremost their currency is advancing human knowledge (measured through NCI, see Figure 1, page x). Universities should not be penalised for not creating startups or jobs, nor should they be rewarded for doing so – it simply isn't their purpose. Universities teach students, some students and some teachers start companies, and some of those are successful. We should celebrate that success and thank the university for training the people in those startups. When industry funds a university to do research, then it isn't funding a startup or employing its own researchers – universities want their graduates to find great jobs in Australia, hence the catch-22. Similarly industry isn't driven by getting a gold star from government; it will hire or fire Australians, and pay or avoid Australian taxes, depending on its business strategy, so measuring inputs is not shifting the needle the way we need it to. Unless job creation is financially rewarded in universities – and I'm not suggesting it should be – then why would universities do it? Even if job creation was to be financially rewarded, you have to measure it first so you know whether your incentives are working. Chair of Industry, Innovation and Science Australia (IISA), Andrew Stevens, agrees, saying, 'We have a national preoccupation with increasing the supply of innovation inputs (e.g. skills, research capabilities and activities). We keep tipping in supply side inputs and yet, our innovation outcomes aren't increasing (or increasing at anywhere near the same rate).'[37]

To be fair, today we do measure some outputs. Some universities do count company creation every year, but the company could be a professor with an idea who files a patent and pays $1000 to register a 'company'. If we measure just company creation and not growth, then it's the wrong metric. CSIRO and others measure outputs like number of patents filed, but when Australia's most prolific filer of patents is consistently a gaming machine manufacturer (in 2022, Aristocrat Technologies filed 71 patent applications, followed by CSIRO with 52 and NewSouth Innovations from the University of New South Wales with 29),[38] is that really what we mean when we say we want Australia to be more innovative? On that output measure, wi-fi was a great success: we filed numerous patents, and have returned royalties to Australian science through numerous court actions, but how many jobs and companies did we create? How has it strengthened our economy?

Jobs, revenues and economic growth produced by our innovation are hard, scary things to measure because they are unpredictable. But if we are going to look 'innovatively' at innovation, we have to get comfortable with a way to measure the jobs and growth that we create domestically. Measuring return on royalties for IP is not actually the outcome we want for an innovation-driven economy if those dollars are not translated into broader benefit. Rather, we must measure what is created here in Australia, so that we can justify how we're spending taxpayer funds.

One of the ways we can consider modelling for this is through understanding the rate of growth for a startup compared with a traditional company. A startup might have 10 employees initially, but what are the revenue projections and what is the wider ecosystem it's enabling through suppliers, vendors, customers and so on? You can't just look at the raw number of jobs in each startup; you also have to consider the network effect that these disruptive companies will have. Disrupters create a phenomenal network effect, and all boats rise on that tide. Once we figure out how to actually quantify that in a way that we all believe in, then we can really 'be innovative' about our innovation metrics by pivoting, tweaking and iterating until we optimise that outcome. If we're brave enough to stare that down and live with the initial criticism that startups might only initially create 10 jobs, we will have changed the expectations and drivers for the system and unleashed its power.

In 2015, the CSIRO Innovation Fund was still a gleam in my eye when we encountered Baraja, a startup working on machine vision for autonomous vehicles using light detection and ranging (LiDAR) technology. My eyesafe laser had enabled LiDAR to be deployed in public back in the 1990s, so I was very keen on Baraja. While CSIRO had no way of providing equity investment at the time, we could offer its team lab space at our collaboration hub in Lindfield, Sydney. Two years later, Main Sequence, the independent venture capital fund founded by CSIRO, invested in them; 4 years later they had outgrown the space we could offer them at Lindfield – no mean feat given it's roughly the size of a large university campus. Baraja has gone on to secure investment from the likes of US VC firm Sequoia and global manufacturer Hitachi,[39] and earned a valuation around $300 million.[40] Where are we capturing these measures of success nationally?

Main Sequence partner Phil Morle said that we need to make whatever we measure more transparent and improve how we promote progress: 'We need to do better at showing dashboards of development. Anyone inside this industry can see it taking strides of progress by the year, but on the outside, it is hard to see. We can't continue to refine what we are not measuring.'[41] By not measuring innovation, not only do we hinder our ability to improve it, but we also prevent others from seeing its value and importance. It's hard to make the case for leaping across the Valley of Death if you can't see the benefits waiting on the other side.

Fear of failure

Factors as broad and difficult to dismantle as culture, history and legislation all have a chilling effect on our appetite for risk in Australia. Board directors in

Australia carry personal liability for failure; employees who leave with IP are seen as posing a competitive threat; companies see stronger return on investment on things they've tried before than investment in unknown R&D; and even when we do invest in venture, we reduce our chances of success through hyper-cautious behaviour. When the risk-averse behaviours of all these actors in the innovation system are added together, they have the same multiplying effect as a system that works together to boost innovation – but in the opposite direction. A collective aversion to risk halts innovation at multiple steps in the innovation process. It's the difference between not crossing the Valley of Death because you can't, compared with not even trying at all.

Fortescue Metals Group Chair Dr Andrew Forrest said there is a broader cultural reaction against people who fly too close to the Sun in Australia:

> We have a quickness to glorify failure, to point out the deficiencies and the faults. When someone has a crack and does fail, it's almost a celebration, certainly in the media. That leads to a cautious approach. The people who will still be visionary, will have to not care much about what other people think. They will have to know that if they do happen to fail, then they'll learn in that process and it will make them stronger. There's no success in success; success comes from how you handle failure. That's kind of lost in the Australian culture, that 'oh, failure is failure; that person won't ever get back up again.' Whereas in North America, failure is seen as a really good thing, provided you're still there and you haven't given up. I think that's what holds Australia back is that quickness to criticise and that aptitude to celebrate failure as opposed to success.[42]

Similarly, former Australian Prime Minister Malcolm Turnbull reflected on how hard it was to build national confidence in a climate of negative traditional and social media, noting:

> ... even when things were fine, we were told we should feel bad. Say it enough and you soon will feel bad. That was disturbing when so much of what we do – as individuals, businesses, nations – depends on confidence. We don't have to delude ourselves with false confidence in Australia – we're so hypercritical of ourselves, so good at knocking, that simply restoring a bit of objective balance can work wonders.[43]

National culture

Nobody likes to fail, but I have found that Australians can't even tolerate failure when you're trying something new for the first time. When I was at venture firms Southern Cross and Blackbird, I quickly realised that whether or not I invested in an Australian founder would depend on how they answered when I asked them to tell me about a time they'd failed. If they painted themselves as the star of the story, despite having failed, I knew they weren't open to failing and learning from those experiences, and so I'd think twice before investing in them. It's amazing how much people are willing to try if you take away their fear of failure – but at the same time, imagine how much more we could try if we accepted that failure would happen at least some of the time. Freelancer.com CEO Matt Barrie recalls his first company, Sensory Networks, haunted him when he started Freelancer.com because it hadn't yet sold to Intel and wasn't performing well. He said: 'For two years, I got asked every time I came to any business meeting, "What happened to Sensory? What happened to Sensory? What happened to Sensory?" It eventually sold to Intel. I can say I atoned for all my sins because it sold.'[44]

By comparison, Dr Josh Makower, who is a successful serial entrepreneur, a Special Partner at Silicon Valley venture firm NEA and Professor of Medicine and BioEngineering at Stanford University School of Medicine and Engineering, said he was heavily influenced as a student at the Massachusetts Institute of Technology (MIT) by one of his professors, Ernesto Blanco. 'I have a million quotes from him, but my favourite one is he'd say, "My career is a series of failures, culminated by success." That was inspirational for me.'[45] Josh said students of his BioDesign class at Stanford regularly ask about failure, and he gives them this advice:

> I think about failure as not whether you have failed, but what you do after you have failed. That defines whether you're going to be successful or not, because failure is just part of the process. It's how we learn from failures, recover and persevere forward that really matters.
>
> I remember my first day as an entrepreneur. I got to the end of the day and I had already made so many mistakes. I asked myself, 'How am I ever going to be good at this if I'm just failing left and right?'.
>
> I realised that failure is just part of the process. My goal from then on was to try to not make the same mistakes twice, and to learn from my mistakes in the hopes of doing a better job next time. There's just no way that you can accomplish new things without stubbing your toe and getting it wrong, a lot. But if you can persevere and learn from those mistakes, you can ultimately win in the end.[46]

Steve Blank, founder of Lean Launchpad and I-Corps, uses a sporting analogy for a sporting nation like Australia, saying:

> That's why, in innovation clusters, venture capitalists allow people to take multiple shots at a goal. Imagine you were playing your football and you only get one shot, well, the game would be over. When I meet

foreign visitors at Stanford, I say, 'We have a special word for a failed entrepreneur in Silicon Valley,' and then I ask them, 'Do you know what it is?' I tell them, 'Experienced. In any other country, it's called a failed entrepreneur'.[47]

Blackbird co-founder Niki Scevak said failure should not only not be feared, but actively encouraged:

> The majority of startups aren't going to raise their growth round because they shouldn't raise the growth round. Every startup being able to raise a growth round is the opposite of healthy and there should be a drop off at each stage of progression. In recent years, the conversion rate from startup to growth company was too high and too unnatural.
>
> Failure is part of the system and healthy because it gets those people to recycle back into the winning companies and it gets them to recycle back into starting new companies. The saddest thing is when companies don't fail or don't succeed, they get stuck in this middle ground. The prime years of those people involved in those startups are being spent in this middle state of flux of not failing, not succeeding. To me, that's the saddest thing.[48]

On the industry side, fear of failure also has a chilling effect on R&D investment. One of the reasons that we don't get much innovation in industry in Australia is because most businesspeople are more focused on the risk and governance side of the equation than they are on the upside and opportunity side. It's easier to get US CEOs and companies to that upside thinking than it is here in Australia. For example, before becoming Director of the ANU's School of Cybernetics, Professor Genevieve Bell was one of the first women Fellows at Intel in Silicon Valley, the first woman to be made a Senior Fellow, and the first woman to hold both the Senior Fellow and VP title:

> We were assessed on our risk-taking skills, not our ability to avoid and manage risks. The expectation was that we were always pushing and that things could and should always be better, generation on generation. Moore's Law embodied that, but it was more than that. It was the logic of the company. And there were lots of values and cultural features that helped support that – constructive confrontation, meritocracy, a rejection of classic management hierarchies, and the idea of disagree and commit.[49]

One of the Australian exceptions I've encountered is Fortescue Metals Group Chair Dr Andrew Forrest. Andrew said Fortescue's 10 corporate values include empowering his people to 'always, not casually, not when you feel like it, always have a crazy brave Plan A. It will likely fail, as really good Plan A's do, so don't proceed without a bulletproof Plan B. Your workforce have to believe they are there to make a difference and they come to work to take risk at every level of the organisation. If it gets up to my level of the organisation, I will always give them cover to take risks.'[50] In Chapter 5, I talk about Andrew's ability to predict how his investment in hydrogen research could have a 'multiplier effect', finding many applications from the initial innovation, which is something we need to see more of in industry. Any researcher will tell you that you can't contract them and pay them $1 million to deliver an invention in 9 months. It doesn't work that way because it's not a linear process. Australian companies have difficulty funding research because they want that kind of clear outcome and don't necessarily see the value of failure in trying to pursue that, but they have a different mindset in the US. An American company might fund research just to find out if someone else could do it, resulting in disruption for them, and actually be happy if the researcher comes back and says it can't be done – failing that key performance indicator (KPI) generates a reward. Three of my startups won support from the US Government's Defense Advanced Research Projects Agency (DARPA), but the wildest one was Translucent. I told the Program Manager we just weren't ready yet for funding, we needed to prove the idea, and he said even proving it didn't work would be a win, so stop arguing, take the money, and get on with it. Where would you hear that in Australia?

Many factors stop us from successfully crossing the Valley of Death, but often fear of failure is enough to prevent us from even trying. In 2022, I gave a speech at *The Age* newspaper's Innovation Summit in Melbourne on the topic, 'Steering Australia through the Valley of Death', where I said:

> Rather than risk the Valley of Death, our best ideas go overseas, and we buy them back once someone else has taken the risk for us – and taken the value from us … When you look at all the company creation programs CSIRO supports for Australia over the past few years, we've helped create more than 250 companies and thousands of new jobs from Australian science. That's 250 times we've leapt over the Valley of Death and made it to the other side – but it only takes a few falls to make us question ourselves again. If we let fear of the fall stop us from backing our own ideas, they will keep going overseas and crossing the Valley there – delivering value and solutions for other countries.[51]

In Chapter 4, I talk about the importance of 'near-death experiences' for a startup in testing themselves and emerging stronger, because if we are too afraid of failure and risk, then we will never get past the first hurdle.

Cautious boards

Australian legislation means board directors here can carry much higher consequences for a company's failure than in the US. We're absolutely world-class when it comes to governance and protecting shareholders and customers from wrongdoing, but these same laws can also have a significant chilling effect on appetite for risk, which is essential for innovation. Everything that works for you in the good governance of a public company can be a nightmare in a startup, certainly if you want to be agile. Australian boards often take a 'trust but verify' approach, asking probing and challenging questions that can help protect the board members from being sued. In the US, directors of a startup's board, who are mostly venture partners, are usually focused on maximising the value of the company, so their actions are more designed to make the company successful than to protect themselves from liability. They'll do anything to help you win. I haven't seen a lot of that mentality on boards here, because ultimately, directors are personally at risk, so they're more likely to put themselves ahead of the company and that's exactly what legislation encourages them to do. Established companies are the most common kind of company in Australia, and so are the chief consideration in drafting legislation, but that legislation doesn't make as much sense when applied to a modern startup. Even worse, because the startup approaches the edge of a Valley of Death pretty much every year, as they raise the next round or struggle for the next big order, I've seen many directors resign to avoid liability.

Startups are always running out of money; they live on the edge of insolvency because that's the most efficient leverage of equity over capital. Australian board directors are caught between a startup's reality and the realities of personal liability and insolvency stringency, which are perfectly logical for an established company. Worrying about cash-flow can distract a CEO by making them chase today's market, rather than getting ahead of the market to lead the market – essentially selling tomorrow to buy today. Of the three possible outcomes of all companies it traps you into the worst: instead of success or failure, it sentences you to purgatory. Earlier in this chapter, Niki Scevak describes purgatory as the 'saddest thing' that can happen in someone's career. So many Australian startups end up as 'lifestyle businesses' that create enough revenue to feed everyone, but never enough to grow. The Australian economy is 99.8 per cent made up of small and medium businesses[52] because so many companies get caught at this level, unable or unwilling to grow further.

I think we need to create space for Australian startup boards to be more like US VC boards when it comes to deep-tech startups. Venture capitalists invest their money, but in Australia, board directors invest their reputation. I think that makes a fundamental difference in risk tolerance. The venture capitalists want you to take risks, because the whole reason you invest in a startup is that you're willing to lose everything in order to win big. But a board here often doesn't really want you to take too much risk, or only a calculated risk, because board members' reputations are on the line. It is a really tangible difference. As a country, if we want startups to be empowered to take risks and have boards who support them, we need to rethink personal liability for directors – regardless of the size, impact and age of a company. It's one thing to have director liability when you have a workforce of thousands and a customer base relying on you for an essential service, but quite another when a company is still getting off the ground and is barely revenue-positive yet. You can't expect boards in this environment to make revolutionary change – or at least, you can't expect a board of a big company to act like a startup board because those dynamics are fundamentally different and that difference really does impede innovation.

Catherine Livingstone has been on the boards of organisations as diverse as CSIRO, Telstra, Commonwealth Bank and startup Saluda Medical, to name just a few of her positions. She said Australian boards do moderate their approach depending on the scale and impact of the company, but that governance rather than mentoring is their focus:

> The principles of governance are not different, but the practice of governance, and the depth to which the Board needs to go to demonstrate governance, because of the people who are providing the capital, [mean you adhere to] core principles but [they have] to be fit for purpose. So if you get a Telstra or a CommBank, and you have regulators and you have shareholders, (retail and institutional), and community really interested because of scale and impact, so you have a whole range of stakeholders and your governance must address that.
>
> Compare that with, I'm on the Board of Saluda Medical, which is an implantable medical device startup that's been going for 13 years, and it's just received FDA [Food and Drug Administration] approval in the US, so the governance has evolved from a few angel investors, then to venture capitalists on the board, so you're managing conflicts of interest, then it'll have to evolve again, because it'll at some point go through IPO [initial public offering].
>
> So the governance has to evolve, but the fundamental principles, like transparency, like managing conflicts of interest, like good systems, reliable financials, having a plan, managing risks, they all have to be there, but in a fit for purpose way.
>
> To come back to the governance point and fit for purpose and the core principles, I think one of the issues around startups is that they may

not establish those core principles from the beginning. And then, particularly on the conflict side, because it's not been established that the conflicts have to be managed, conflicts come up, inevitably, when you have a small number of shareholders, and a founder, and there's no mechanism to deal with them. It's then you get the implosion.

Whereas if you, from day one, when you have nothing, if you establish those core governance principles, they'll stand you in good stead.[53]

Boards often see risk where entrepreneurs see opportunity; boards usually want systems and processes whereas entrepreneurs don't ever want to do the same thing twice – they want to do it differently every time to make it better. The nature of young companies and startups is that they keep trying different ways to do things that big companies have taken for granted, even things like human resource systems or safety systems. They invent it all like it's the first time and, because they have that mindset, they tend to build far more efficient systems than big companies do.

The other thing that Australian boards want is often a full-blown business plan. They usually want really clear KPIs, metrics and milestones that they can measure. When I ran venture-backed startups, I was never ever asked for my business plan. They'd ask what the idea is, they'd ask to see the expected 10 PowerPoint slides (no more!) for your pitch, they'd ask how you were going to do it, and then they'd critique it and you'd go back and forth – but we never wrote a business plan, even when we got the money. The first round of funding always has the same milestone of proving the idea really works – it's not that complicated. You're not using a Gantt chart to map out all the steps. Everyone in the company is just obsessed with that one goal. I've seen large Australian companies attempt to deploy 'agile' methodologies, following the 'lean launchpad' approach, and if you've run a startup you'll know how extraordinarily badly 'agile' can be interpreted by some of corporate Australia.

Getting a board to sign off on even a traditional investment can be a challenge – it's designed to be that way – so imagine the challenge in seeking board approval to invest in a market that doesn't even exist yet, such as hydrogen fuel. A traditional investment analysis of a market uses a net present value (NPV) assessment of the existing market, or at least there are parallels to them that you can analyse. For our hydrogen example, if you wanted to invest $20 million, the board would ask how big that market is, but there is no market for hydrogen fuel – yet. They'd ask what the NPV on the $20 million would be, but it's not a piece of manufacturing equipment – you want to use the money to fund something that hasn't been invented yet. Essentially you're telling your board that you want to buy stock in a company that doesn't exist, and you're hoping that company will make a breakthrough and create the product. I think you can guess what their response is likely to be. One of the reasons US venture is so ridiculously successful is this willingness to take risk that goes way beyond conventional governance frameworks, especially ours in Australia. Great venture creates a whole new market, so you have no comparison or conventional analysis tools. When plant-based meat alternative company Impossible Foods started, there was no market for what it was doing, but the company created it. Optical communications had no market, but that technology helped build the internet.

CEO of Australia's Future Fund Raphael Arndt said the Fund had to convince its board to endorse its expansion into VC in the peak of the 2007–08 global financial crisis, which, he said, 'wasn't hard – but you had to come at it from the right direction':

> We had to paint the picture of the opportunity, and then we had to build a process that still had appropriate checks and balances because, after all, we're investing taxpayers' money. A big part of it was changing the process so that the board and the investment committee approved the strategy, and not overarching capital allocation. So, the individual opportunities were able to be approved by the team with just one other person, which was typically me, in those days, signing off, and I'd guarantee that I'd do that within 24 hours.[54]

Maile Carnegie, the former Google Australia Managing Director, moved to ANZ Bank to lead its digital transformation program in 2016. When I spoke to her, she was four years into that process and, by her own admission, not even halfway. But what I found gutsy was ANZ's willingness to announce their clear commitment to digital transformation before they had a plan in place – to lead with their goal first. So often in the corporate world and in governance, you wouldn't allow that to happen. But that's the way a startup would do it – you'd say we're going to solve this problem, we have no idea how we're going to do it, but we're going to figure it out. When ANZ announced the transformation of its retail bank, there was a lot of noise in the market and a lot of criticism in the media. But what a gutsy thing to do. I'd argue that only someone who came from a deep-tech company like Google could bring that mindset to a bank and overcome the traditional, risk averse governance.

For all the extra layers of governance, I will say I have learned the value of the Australian practice for the Chair and the CEO to be two different people. In the US, those roles are often filled by the same person, but two brains are always better than one when it comes to innovation, and the more diverse the views, the better. A good Chair can sit back and look at the big picture of what's going on, while the CEO is buried in operational matters. I learned this as Managing Director of Arasor, which is discussed further throughout this book. If you have the right appetite for risk, then this need for additional scrutiny can be an Australian advantage.

Better Place Australia CEO Evan Thornley and CTO Dr Alan Finkel say boards can be a blessing or a curse depending on the makeup of its members. When the global parent Better Place company was starting to wobble, Alan said its board barely noticed because the board members were 'all representatives of the investors who knew nothing about the business and didn't add any value and didn't even see it failing. We could see it from inside much faster than them. You can't really blame them; they had been parachuted in irresponsibly by the investors who thought it was all easy.' Evan said ineffective boards are the number two reason startups fail (the number one reason being running out of cash), brought down by disputes between founders, between founders and venture capitalists investing in the company, or between venture capitalists from different funds investing in the company. Conversely, Alan said he deeply valued his board when he founded medical device company Axon Instruments and had a small board of industry experts and experienced business leaders. He said the key difference is how people end up on your board:

> Boards fail when people are representing the investors, and investors haven't got a clue who to put in to represent them, as opposed to a skills-based board. It doesn't matter where the money is coming from, you should have a skills-based board. As CEO, if you've got a big investor who wants to have a representative, you really need to be on the front foot and work with that investor to say, 'Fine, you can have two board seats, but can we work through the skills matrix work with you and find the people who can represent you while actually adding value?'[55]

That's the model we should be encouraging for deep-tech startups. It is a fundamentally different way of running a company and our failure to understand that is one of Australia's impediments to solving our innovation dilemma.

Unadventurous venture

As a nation, we invest less than the OECD in research but our universities still perform in the global top 10. Our capital market – the fourth largest in the world[56] – is not investing in this world-class research to deliver world-class innovation; that's our second Valley of Death discussed in the Introduction (see Figure 2, page xi). I believe it's partly due to a fear of failure in our culture, which translates across to the legislation governing our corporate boards, then also filters through to our nascent venture system – the last place it should be. Australia's history of venture is riddled with stories of investors who were too risk-averse to invest until they saw everyone else investing, but by that time an investor has lost their competitive advantage. Blackbird co-founder Niki Scevak said that after Blackbird's first failure (a company called Ninja Blocks, discussed in Chapter 2), he embraced failure as an inevitable part of venture, easier said than done in Australia:

> Everyone outwardly says that most startups fail. That's easy to say on a panel or in a blog, but it's so hard to do. The psychology of a successful investor is in training the muscles in your brain that actually, there isn't a lesson to be learned by a single failure. It's going to be embarrassing and it hurts and telling people they've lost all their money is a horrible feeling. But at the end of the day, that failure is not a valid signal until you get a large enough sample size. If there's 20 failures in a row, then that is telling you to change your behaviour. But if startups are a 1 in 10 chance of being a great success, then you want to have enough of a sample of investments for those low probability events to be able to play out or not play out.
>
> We had the mindset that, absolutely, things were going to fail. I would almost say the problem of the first [Blackbird] fund was that it didn't fail enough. Too many of those companies – particularly SaaS [Software as a Service] companies – there was an almost 100 per cent hit rate of creating a business and being successful after the first investment we made. They're all different sizes and so on, but that first investment actually didn't fail enough. So, looking back, were we actually taking enough risk?
>
> If you're going to do venture capital, you should lean into risk. If you're going to do venture capital and take a risk-minimisation approach, then you're in the wrong business. It's an ugly combination, because there are so many startups that would love to have some money invested in their company, so you'll always find someone to take the money, but when you combine this risk aversion with the ultimate risk asset class, you're in the middle of the road and you're going to get run over.
>
> Horsley Bridge found the worst venture managers they invested in failed 50 per cent of the time and then the best venture managers they invested in failed 60 per cent of the time. [The better venture managers] took more risk and did something fresh because venture capital is about doing something new. It can't be part of a trend, it can't be the 40th whatever, it can't be successful in the US and you're doing it from Australia. It has to be something fresh and original and new and so we're always very mindful to make sure that we're failing the right amount of time.[57]

Former Southern Cross venture partner Tristen Langley characterises Australia's risk appetite in venture:

> You have to have that perfect storm, and there just isn't enough dust being kicked up in Australia to make the storm happen. The perfect storm is great innovation, meets an entrepreneurial attitude that can push it out into the market, meets an investor that believes in it, meets engineers that can help build the product even more. If you kick up enough dust, you can actually get the storm to happen. But Aussies say, 'I'll dip my toe in here,' and then just as quickly, 'Ah no, I don't think now is the right time.' They go around in circles so things aren't coming together. You need some real momentum.[58]

I discuss the waves of venture in Australia in Chapter 2, where you can see a consistent theme where funds tried to de-risk their investments by making safe decisions, when actually a safe decision often becomes an unsafe investment because you've lost your early-mover advantage. Each time funds failed, they made the next round of investors more risk-averse and therefore less likely to invest well. I've heard stories from Australian investors who pat themselves on the back for not investing in 'stupid mistakes', but they don't have any runaway successes to show for their risk aversion, either.

Silos and cultural cringe

When I came home to Australia to be Chief Executive of CSIRO in 2015, I was on a panel and during the discussion, the moderator referred to me as 'just a scientist'. To me, their implication was clear: as a scientist, what would you know about business? Reviewing some of the television interviews given by my predecessor, Dr Megan Clark, during her time as CEO, I was shocked to hear her called 'head of the nerds'. She laughed it off, but I don't think I could have. After 26 years in the US, I'd forgotten what Australian culture thinks of scientists. Sure, they're smart, they're achievers and they're important — but don't ask them about anything other than science. We need to embrace a more diverse model for success if we're going to be competitive in an exponentially changing world. People are so much more than the sum of their qualifications and it's one-dimensional thinking to try to label them or confine them to a box. Not only is that absurdly narrow-minded; it's also hugely damaging for the next generation of STEM specialists – kids in school who see that and like science but don't want to be called a nerd. We need our kids to be empowered to change their world.[59]

To be fair, as former CSIRO Chair Simon McKeon points out, we're bad at letting people out of their boxes in general – it's not just scientists we confine to their areas of expertise:

> Unlike other countries, we are still prone to be very relaxed in our silos, whether it's academia, business, government, even the not-for-profit sector. It's not as if there aren't some that cross-pollinate and transfer from one to the other, but we struggle at really, genuinely trying to mix it all up like other places do. For example, in the US when a new president is elected, there's literally a fresh bunch of businesspeople that come into government and hopefully do good and positive things. A new administration brings with it fresh outside thinking, not just dinners and lunches where politicians rub shoulders, but they actually become part of the machine. The relationship between the American university system and business is even more well-known. We're conditioned to go into one of those silos early out of university and there's not a lot of people who not only actively moved from one to the other, but are even encouraged. We just don't have a culture of mixing it up like other places. I think that comes at considerable cost for this country – there's just not enough shared experience.[60]

As frustrating as I've found this Australian fixation with putting people in boxes based on their education or training, I wonder if it explains the cultural cringe we have around certain fields that require expertise you can't pick up in traditional education paths. Since coming back to Australia, there are two words in particular that I've noticed set off unhelpful reactions, so we probably need new Australian words for them to better define what we mean. The first is 'entrepreneur'. I was the first Australian founder to be backed by US venture fund Accel when it was a rising star in venture, well before it invested in Atlassian and many other companies that went on to become phenomenal success stories. Accel's partners were intrigued by this trickle of Australian founders coming to the US with great ideas and with no venture system back home to support them. Their intrigue turned to horror when Australian scientists wanting to be founders arrived in the Valley from Australia in the 1990s and 2000s and brought 'advisors' with them to make up for their own lack of business experience, which I discuss more in Chapter 2. These advisors would try to run the pitch with slick presentations, plans and business cases, which the scientists were all too happy for them to do because they couldn't fathom ever being an entrepreneur themselves.

I think even today Australians – not to mention Australian scientists – cringe over the word 'entrepreneur' and directly associate it with a get-rich-quick scam. Tristen Langley recalls meeting would-be scientist–entrepreneurs in Sydney in the early 2010s at tech-networking events:

> What you noticed that was different at Australian events compared with in the US was no one really knew how to fundraise. You would be having a great conversation about the idea or the technology, and then one of the team would say to another member of their team, 'Are you going to ask for the money?' And the other would say, 'No! You ask for the money!' No one really wanted to step up and do the big ask.[61]

Australian scientists tend to think of the worst kinds of business leaders because, more often than not, those are the ones in the headlines. You don't see many headlines about good leaders doing humble things because that's not 'newsworthy' today, and as a result it's a hidden model for leadership. Scientists aren't usually getting *Harvard Business Review* delivered with their *Nature* subscription, so they make assumptions about entrepreneurialism.

CEO of the Australian Future Fund Raphael Arndt has overseen the growth of that fund's investment into deep-tech venture, and said Australian scientists:

> have often been less inclined than in the US to be flexible when it comes to funding and finance models. They tend to say, 'I'm a pure scientist and that's it,' and to see private capital as an evil, but actually, bringing applied science into the frame and obviously, that's what CSIRO focuses on, actually helps good ideas to propagate and get picked up and make a difference to people's lives. And that's a good thing. So I think there's more to go in terms of what we can do there.[62]

Perhaps that's why scientists often go on to become the kinds of CEOs who are more naturally humble, more grateful to the people who got them where they are and more likely to share more of the wealth of the company, because they succeed through the additive nature of science-driven innovation, not innovation that steals market share from others (I talk more about this in Chapter 3). Unsurprisingly, my arrival within CSIRO as the 'entrepreneur' was met with scepticism by some who thought I would have a commercialisation agenda for every single project.

It's hard to train this cringe out of Australian scientists when it's also widespread in our culture. Matt Barrie remembers that, just before he made his final – and successful – bid to buy Freelancer.com, his mother told him to 'get a real job' after his previous entrepreneurial endeavour, Sensory Networks, hadn't gone as well as hoped.[63] Founder and CEO of Spreets Dean McEvoy was back at home, broke, and still developing the idea for Spreets when his parents told him to 'get a real job' and pay them back the money they'd lent him as an entrepreneur: 'It was hard to argue with them. I think the prospect of having to go and get a job was so miserable that I just decided to find a way to not have to do that, because I felt like that was totally selling out on my vision and purpose as a person.'[64]

Steve Blank, founder of I-Corps and the Lean Launchpad methodology, credits US academic AnnaLee Saxenian for identifying the difference in culture as a key distinction between the performance of innovation in Silicon Valley/San Francisco CA, versus Route 128, Boston MA, in her book *Regional Advantage*:

> Saxenian said if you started a company in Boston, you were probably living within driving distance of your parents' house, so you probably had dinner there often. Your Dad – and it was Dad at that time – worked for the same company for 30 years and if you said, 'I just changed my job,' they'd hang black crepe in the window and say, 'Don't tell that to the neighbours!'.
>
> But most people had to come out to California, pre-internet, for not only different work culture, but remember it was next to San Francisco, so they could have different lifestyles as well that they didn't want their parents to know about. So the culture in Silicon Valley, and the great universities, allowed a very different set of risk culture, away from parental norms. Saxenian made me realise, 'Holy cow, culture matters, risk matters, risk tolerance matters, capital at scale matters, university matters.'[65]

Australian Tech Council Chair Robyn Denholm describes Australia's inherent conservatism in career choices:

> 30 years of uninterrupted growth hasn't created a culture of taking risks or changing the status quo. This is as true in the boardroom of ASX companies as it is in government and in our society. Why do we produce so many great lawyers and accountants? They are viewed as the 'safe professions'.[66]

Unfortunately, I still know of investors and lawyers today who have this attitude towards founders – naive deep-tech founders in particular – and do their best to take advantage of them through term sheets and other arrangements.

The second cringe I've become especially aware of in recent years is around the word 'innovation'. One of the most common reactions I encountered when I started to talk about innovation after taking on the role at CSIRO was a fear of automation and digital technologies taking jobs. This surprised me because prime ministers going back to Billy Hughes in the 1910s and probably earlier had talked about science and innovation as essential to our economy. But as discussed throughout this book, I came to realise innovation was largely correlated with digital technologies (as opposed to other deep-tech innovations) and that these themselves were not well-understood in the wider community. The perception was that technology was coming for our jobs, and that we would not succeed in harnessing them to create new industries and jobs to replace them.

Former CSIRO Chair David Thodey doesn't think Australia's innovation aversion comes from fear of failure; he thinks it comes from complacency and lack of awareness:

> I'm not convinced that Australians fear failure. Look at mining and agriculture, the risk that farmers take every year and the incredible geological work we've done as a nation. These are all high-risk initiatives that have a reasonably high probability of failure. So I don't think that Australians are inherently more fearful of failure than others. It is true we haven't been as invested in the technology sector, and I think that is because there have been better returns in other industries for Australia. However, that has caused our economy to lack complexity – being focused on a few large and very successful industries. In that process, I think we have missed an opportunity, because we could have a great resources sector, a great agrarian base, and a wonderful technology sector. So, I see this through a lost opportunity lens, but the opportunity to build a stronger information services and technology sector is still available to us. We need to talk about the exciting possibilities and have a culture that celebrates innovation. If only more Australians knew some of our talented scientists and innovators in Australia as well as they do our favourite sportspeople – that would be an encouraging start. So I don't think our aversion to innovation is due to fear of failure, I think it is driven by other factors.[67]

Similarly, CSIRO's Chair before David, Simon McKeon, contrasts Australia's complacency bred from good luck with Israel's pride in innovation and connects it with national narratives:

> Australia has pockets of absolute world's best brilliance – but on the other hand, we have had challenges for a long time that we, as a whole nation, don't seem all that interested in changing. We're a lucky country, I think the luckiest country. Everything seems to fall our way. That can, of course, lead to complacency. Luck means we don't have to try as hard as other places. With the most sophisticated social security system in the world, it's easy to sit back and watch others achieve as a nation. It might just sound totally naive at this point, but it does require a message, a narrative, a 'this is what we can be' consistently coming from our leaders, and at the end of the day, we have never had that consistent message.
>
> Also, it's really important to focus beyond the celebrities of innovation. We all know that behind every celebrity is a really important team. There's huge need to tell the stories of those really talented people more prominently than we have. You see more innovation in countries that celebrate the second and the third and the fourth lines of those extraordinary machines that grew, not just the celebrity at the top. Take someone like Craig Winkler, former CEO and largest shareholder in MYOB, whose natural disposition was to be a highly

effective leader but someone who didn't need to be the 'loudest voice'. He has gone on to invest in other innovation companies like Xero and sit on their board, guiding, helping, and in his own very understated way. I bump into more Craig Winklers than Mike Cannon-Brookeses [high profile CEOs].

One of my memorable experiences was being in a taxi in Israel driving through a new industrial estate lined with Microsoft and Apple premises and other tech names out of the US. The driver said with great pride, 'This is the new Israel – and I'm part of it. You wouldn't be able to have your meeting if I wasn't driving you there.' It was spot on. That culture just spread right through Israel. He was genuinely turned on by the opportunity it gave him, even as a taxi driver. I've never heard that in Australia.[68]

Conversely, and perhaps reflecting his role in shaping the national innovation narrative, Emeritus Professor Roy Green, Special Innovation Advisor at the University of Technology, Sydney, said while the innovation challenge is 'certainly not one that can be fixed easily', he does think 'the situation has improved because we've just been talking about it relentlessly'.[69] Fellow academic Distinguished Professor Genevieve Bell, Director of the ANU's School of Cybernetics, presents a counterpoint view: 'We aren't bad at innovation here in Australia; we are bad at telling stories about our innovations and that is a hugely important difference. [It is damaging] culturally and politically to say, so consistently, that we are bad at it.'[70]

This cultural cringe makes it hard to build broader cultural engagement with – and advocacy of – crossing the Valley of Death. Commercialisation is not seen as a matter of public good, regardless of the challenges it solves or the jobs it creates. Main Sequence partner and ON mentor Phil Morle said our culture does hold us back – but it's up to entrepreneurs to push through:

Mindset has been the biggest challenge to conventional thinking. It is mindset that has historically held Australian innovation back. Fear, over-thinking, greed, secrecy, tall-poppy syndrome ... these things have created a system that holds on so tightly that nothing gets done. Building a culture around execution, courage, risk, magic and radical collaboration leads to remarkable leaps of impact and value that the system is starting to discover.

I convince people by doing it anyway. There's that old saying 'If you don't believe it can be done, get out of the way.' We had that in the early days of [Australian deep-tech startup] Samsara where we had the idea for a company that would use enzymes to infinitely recycle plastic. Initially people argued that it could not be done. We found a scientist who could do it and we built the company. Two years later this is a $100m+ company.[71]

I do have hope that we can overcome this cultural cringe based on my lived experience in the expat community in the US for many years. The 'Aussie Mafia' entrepreneur community in the Valley – with a lot of sweat from the members and a little funding support from CSIRO – has now grown to more than 300 startups whose members are all there to hear about what each other is doing and help push the 'Team Australia' effort forwards. There is a bit of healthy competition there, of course, but it's in aid of making each other stronger to create a tide that lifts all boats, not competing in a 'finite pie' mindset for a fixed funding pool. They are very connected in a way that they'd never been when they were in Australia, where they had struggled to find a like-minded community. In the same way, the founder network of mentors that we built up around ON (discussed in Chapter 6) is still huge and very active, regardless of what phase the ON program is in.

Perhaps part of the reason our startups would rather seek success overseas than try the leap across Australia's Valley of Death is because we still frame the notion of 'success' as something bestowed by international peers. There is something in our cultural identity that shies away from certain types of success. You see it with the actors or musicians who have to 'make it' in Hollywood or Nashville, or in the business leaders who leave Australia to go on to international success like Charlie Bell heading up McDonald's or Andrew Liveris taking on Dupont/Dow. It strikes me that only after they are recognised internationally are we proud to claim them as Australians, not if they win only Australian film or music awards or if they're CEOs of Australian companies. It's wonderful to celebrate international success, but we cringe over the idea of success at home; this is an active and dangerous disincentive.

More recently I've met members of another Aussie entrepreneur or founder group, one that has come the opposite way: returned expats who have bypassed the local ecosystem but remain connected to Silicon Valley. They are hard to find but well connected to both ON and the Silicon Valley expat group – we even have a few of them hidden on CSIRO sites like Lindfield in Sydney. Each of them has overcome our cultural cringe – these are the people who give me faith it's possible.

Invention, not innovation
We lead with science
Australia has world-class science. Australia's share of the world's scientific publications has been growing steadily and its share of the top percentiles of

highly cited publications reflects the quality of Australia's science and research output. Our academics publish in the best journals, in high volumes, have high citation rates and compare very favourably with our OECD colleagues. For example, in 2022, Australia maintained 10th place[72] for the fourth year in a row in the *Nature* tables, a ranking of citation frequency across all the prestigious *Nature* science journal titles. These are metrics we study closely and we measure return on investment to our universities on this basis.

In Australia, when we say 'innovation system', we really mean 'research system', 'invention system' or 'science system'. We mean: what are we going to invent and publish? But performing world-class research is not the same as doing innovation and we should stop assuming a causative relationship will follow. This is because world-class science should lead with discovery questions whereas world-class innovation has to lead with market vision (discussed in more detail in Chapter 5). Australia's scientists are trained to think that once they've invented something or had it published in a research journal, their job is done and they move onto the next discovery. The hard part is taking that brilliant invention and giving it to a customer who will derive benefit from it and assign value to it. What is often absent is the expectation – or, more critically, the pathway – to turn it into something real.

It's not just in universities, either; we often set our national science priorities around areas of promising research, when national need is the more critical lens – where a market will pull solutions from science. We ask ourselves what areas of research we are going to fund in our universities, like artificial intelligence (AI) or quantum technologies, instead of asking in what markets can Australia be a world leader if we invest in them. We decide we will invest in AI and quantum technologies, but not more specifically AI for agriculture or quantum technologies for mineral detection, for example.

In my first few months heading up CSIRO in 2015, I was invited to attend a meeting of the Prime Minister's Science Council. The topic for discussion was national science priorities and as the talk moved around the table, then-Prime Minister Tony Abbott asked me what I thought Australia's science priorities should be. 'I think you're starting with the wrong question,' I replied.

He looked at me curiously and asked what I meant. I said, 'I think you should always start with the market for the science output, or the challenge you're trying to solve, and then decide what science will help you to do that. It sounds to me like you're starting at the wrong end of the problem, kind of the way some scientists do – falling in love with their research and then looking for a problem for it to solve.' He laughed and said, 'Assuming I am asking the right question, do you think these are the right science priorities?'

I thought about it and said: 'Based on what I know about the market and the challenges you're trying to solve, they probably are the right priorities, but I also suspect it won't matter. I'm not trying to avoid the question, but let me answer it with a specific example. You've listed cybersecurity as a science priority, but I think the priority is actually data and data science, because cybersecurity is just one way of protecting data, so the priority shouldn't be the fence, it should be what the fence is guarding – let's first make our data more valuable so it's worth protecting.' I think he liked that idea.

We don't pick winners

For me, this conversation came to epitomise Australia's approach to science. After my first Prime Minister's Science Council, CSIRO tried very hard to come up with national science priorities that reflected market opportunities for Australia and that would solve some of our greatest challenges. The harder we pushed and tried to focus, the more the system pushed back and broadened again. That's the politics of research – if you say Australia is brilliant at cardiac science, the system will arc up and say 'No, we're good at all health, it's unfair to just pick that one.' It's part of the 'every child gets a prize' philosophy. I was never successful in getting the science priorities shifted into market priorities for good reason, because, politically, you do have to fund everything.

Government policies try to get industry to lean in more, be less risk-averse and fund more science, and simultaneously to get research to be more focused on what industry wants. In 30 years of reviews of our innovation system, there is a strong history of recommendations that 'government shouldn't pick winners'. Certainly, government shouldn't be picking individual companies – that's not how a capitalist nation works – but it's a mistake to interpret this as 'government shouldn't pick markets'. A critical part of innovation is starting with the market and identifying the problem you're going to invest in solving as a nation. Governments can pick areas where nationally we can have a competitive advantage. You have to start from that point and work backwards to figure out what areas you are going to invest in and thereby figure out areas of science to develop to solve those problems. It's never just a job for government, though. Not 'picking winners' has been used as a catch-all to excuse government from intervening too much. Special Innovation Advisor at the University of

Technology, Sydney, Emeritus Professor Roy Green said it's essential we 'pick winners':

> Every country in the world that's been successful in science and technology-led commercial activity picks winners. They might not be companies, but they will be areas of technology in which they can invest. Part of that is recognising that we can't be excellent at everything, we've got to pick winners. The problem is that we spend a lot of money, and a lot of it has been spent picking losers.[73]

Making 'targeted bets' is in line with MBA best practice, which, as discussed in Chapter 4, isn't always the best approach for running individual startups. Managing Director of Microsoft Australia and New Zealand Steven Worrall said if we were going to run Australia like a business, we'd make 'more targeted bets' when it comes to new industries such as quantum computing, hydrogen, solar, battery technology and space:

> If you were the CEO of the country, you'd probably make some more targeted and stronger bets as opposed to letting the market make assessments – because Australia is often not the winner in new technology areas. We might need to reconsider where we have deep competitive advantage, maybe going back to some of [Michael E.] Porter's *Competitive Strategy* and the conversations you might have if you do an MBA. What's your structural competitive advantage? And how are you thinking about your position geopolitically and your role in the world?[74]

This is reflected in the statistics cited earlier in this chapter, which showed that government investment in R&D has shifted from investment in focused areas of research to be more industry-directed. This has shifted the job of 'winner picking' to industry – except that we also know from the data that industry has been leaning less and less into R&D. A critical part of innovation – and a reason why those 30 years of reviews never cracked this – is that for innovation to work, you've got to start with the market. You've got to identify the problem. You've got to pick winners.

It's a big problem because on one side you've got great research, academic achievements and high NCI (normalised citation impact), and all those wonderful metrics with journals, and on the other side you've got industry building and selling products, satisfying customers and managing their business (the culture-driven first Valley of Death, see Figure 1, page x). Even when incentives are introduced to drive collaboration, both sides are talking about the three different 'horizons' of science, where what's called 'Horizon One' refers to the science that's closest to real-world applications (i.e. 'one horizon away'), 'Horizon Two' requires a bit of engineering to be applied ('two horizons away') and the more distant 'Horizon Three' refers to brand new breakthroughs in their fields that are a long way from application ('three horizons away'). The research sector chiefly wants to do 'Horizon Three' research, which is where all the great papers and publications are, whereas industry wants to do 'Horizon One' applications of the research – although in truth, many Australian businesses don't want to do research at all; they only want to know what the products are and then run with them. We call it R&D but there isn't much 'D' going on. It is a fundamental market failure all around the world, but in Australia it's more extreme than in many other countries, both in the science side's fixation on publishing metrics and industry's disinterest in investing in R&D. It's why our Valley of Death has so few bridges, while many other nations have multi-lane highways across their Valleys of Death that make it a narrower divide to cross. I talk more about the need for 'market vision' to guide our science in Chapter 5. Ultimately, I don't think it works to just do the great science and then figure out your market vision later. That's the paradigm that Australia is stuck in now, but we can get better at targeting the problem and being really bold in how we do it.

Finite pie

Australia's small population is spread across vast geographical distances, condensed into city centres by settlers who had to provide everything for their state's population rather than rely on slow and expensive interstate trade. This was perpetuated for nearly a century before Federation began the slow unpicking of state parochialism in 1901. Less than a century later, interstate transport networks became more cost-effective and accessible at around the same pace as international transport networks, soon followed by the digital technologies that now connect Australian cities to each other and the world. As a result, our state-focused systems never rewired themselves into nationally focused systems before they leapt into joining the global community.

At the same time as this 'tyranny of distance' was being overcome with transport and digital infrastructure, the financial largesse of the start of the mining boom in the early 2000s[75] removed any financial imperative for us to look closely at how we would drive our national economy in the global marketplace we were now more connected to than ever before. And so today, each state wants to be the best in the country – if not the world – in many more fields than it can possibly have a critical mass of researchers and facilities for in a country with such a small population. Competing for this title is not just a matter of state pride and economic opportunity; it also positions states for federal funding to progress these ambitions, where they find natural allies in their universities, which also rely on and compete for federal funding. As much as I believe science has the power to 'grow the pie' and create new value, I appreciate the realities of a finite federal budget and understand the power of these historic motivations.

Competition is important and healthy in a thriving research and innovation system, but competition only serves a purpose if someone eventually wins the resources and grows the benefits for everyone by reinvesting in a thriving, competitive system. However, in Australia the people allocating funding in research do not like to 'pick winners' and so it is stretched thin across recipients, with enough to keep programs afloat, but not enough to make it easy for them to push through and excel. I saw this in the early 2000s when I met with different state governments to talk about innovation as part of the US expats tour (discussed more in Chapter 2) and I was saddened to see it again nearly two decades later as CSIRO's Chief Executive. The federal response to bushfire preparedness and resilience after the 2019 Black Summer bushfires was met with competing and conflicting state-based programs (never mind that fires don't respect state borders), and we saw the same in Australia's response to COVID-19, which led to an eruption of competition for vaccine development facilities to be built in various states. Imagine, instead, an Australia where each state thinks like a startup, focusing on its own unique competitive advantage – and, ideally, in a way that leverages each of the other states' unique advantages in a networked approach so that instead of six competitive silos adding up to less than six, they collaborated and added up to 36 (using Metcalfe's law: a network's impact equals the square of the number of nodes in the network).

Across the country, we do have amazing silos of excellence, but we would really rather compete with each other than compete with the world. We don't seem interested in what our peers are doing in our own country, but we're very interested in what people are doing overseas. The only way we grow the 'pie' of resources – be it funding, market share, talent pool or others – is by recognising that we can grow the pie bigger and faster together than we can separately. 'Finite pie' mentality means we'd rather see our local competitors within Australia fail or go overseas than work together to build a bridge that sees more of our ideas across the Valley of Death and grows a stronger economy here.

Australia's future innovation playbook

Australia narrowly dodged a COVID-19 recession and, at the time of writing, is now facing rising interest rates, increasing cost of living pressures, decreasing housing availability, and falling stock market confidence. These are not conditions that typically inspire Australians to take risks – but there's nothing like a crisis to snap us back into action, as we saw in Australia's response to the pandemic. Science organisations across Australia, including CSIRO, were prepared to respond when the novel coronavirus arrived here in January 2020, contributing to vaccine development, testing and manufacturing; research into the virus itself; support for pivoting local manufacturing to make personal protective equipment (PPE); testing of wastewater systems to pinpoint outbreaks; analysis of big data to predict trends; and numerous other responses. Australia's efforts to find a vaccine for COVID-19, 'flatten the curve' and contain the disease sparked unprecedented levels of collaboration between research, industry, government and communities all working in their own way to achieve a common goal. Announcing CSIRO's *Our Future World* report at the Press Club in July 2022, I said:

> Trust in institutions has been falling for over a decade. In Australia, COVID-19 saw trust in scientists spike, but the uncomfortable truth is it took a pandemic for the nation to look to science for solutions. Trust in science led Australia's response to COVID-19, and we can build on that trust now to put science at the centre of leading a united response to the challenges ahead.[76]

If we can harness this same level of collaboration and goodwill to drive an innovation-led recovery and build future resilience, we can create new jobs, grow our economy and strengthen the foundations for the future – a future where our children have rewarding and sustainable jobs, in unique and resilient industries, that secure Australia's wellbeing and prosperity for generations to come. Collaboration is going to be more important than ever if economic conditions continue to tighten, following the short, sharp slump during the pandemic and heading into a period of increasing cost of living, increasing energy prices and interest rates, and, something we haven't seen in years, soaring inflation. We've seen strong growth in venture in Australia in recent years because interest rates were so low that traditional investments simply weren't giving strong returns, so investors were looking for something really different, really unique where they could get some scale, so they took a chance, but that could be all about to change.

In every recession, every revolution and every major shift of an economy around the world, science has created new industries that have emerged from the turmoil, and those new industries created new value that grew the economy. Businesses that invest in research are more productive, with the turnover growth for high R&D intensity firms between 5.9 and 7.3 times higher than low R&D intensity firms.[77] Those science-enabled industries created the jobs of the future. That's the power of science. Science innovation is different to other definitions of innovation – it creates new value that grows the economy by delivering the highest value, not lowest price; it's literally the gift that keeps on giving.

These were the kinds of conversations happening in Silicon Valley when I first arrived in the late 1980s, talking about how to extract more value from the research happening around the corner at Stanford and other universities. I was trying to start these sorts of conversations when I arrived at CSIRO in 2015, but there wasn't the appetite or urgency then that we experienced in the wake of COVID-19. It tells me that we're probably about 30 years behind the folks in Silicon Valley, but we don't want to copy their playbook or go through the same evolution as them, for the range of reasons we will discuss throughout this book.

At CSIRO, we regularly talk to national science agencies in other countries. Canada is often considered to have one of the world's best rates of science and tech R&D take-up in their companies, with roughly 10 times the number of their companies using R&D than Australian companies. Lately though, the Canadian national conversation has been concerned that their companies are not keeping pace with global innovation – and if Canada is worried, we should be really worried. Similarly, the National Science Foundation (NSF) in the US is completely rethinking its approach to innovation because it's worried about the decline in investment in deep-tech startups and innovation as Silicon Valley moves more towards digital and internet investments. The NSF runs I-Corps, a similar program to CSIRO's ON Accelerator program, designed to help science move from benchtop to buyer by teaching scientists entrepreneurship, and discussed further in Chapter 6. So that tells you they haven't been without innovation investment.

Like every other Western country, the US is also extremely worried because of its declining rates of STEM enrolment, whereas other countries like China have an increasing STEM pipeline, and the future of tech depends on that pipeline. Now, if Canada is a leader, not a laggard; and the US, probably the world's best in innovation, thinks it's not performing, then this should be a massive wake-up call for Australia. I'm worried about a world where the US, Canada and China double down on innovation, because I don't see where

Australia fits in that world. There are opportunities for us to work with each of those innovation systems, especially because Australia has such a large and diverse cohort of international students who we should consider as part of our broader innovation ecosystem. But we should do it in a way that also marks out our own distinctive opportunities.

To overcome our national barriers to innovation, we will need to take a uniquely Australian approach that harnesses our strengths, not one that sets out to copy others. This chapter has outlined a number of barriers to Australia truly realising its innovation potential, but the wider book is full of reasons to hope we can change decades of innovation apathy. We can try lots of new ideas and we can push many different levers, but we have to be motivated by something more powerful than economics to make change this significant. I believe we have that motivation – it's what brought me home after 26 years in the Valley. I missed life in Australia. I wanted my kids to have part of their childhood here, and I want them to have as many career opportunities here as they would overseas; I know a lot of expats who feel the same. We shouldn't try to recreate the innovation hubs of Silicon Valley or Israel here, but we can build our own versions by harnessing our love for where we live and a passion for wanting to do what we love in the place we love. This passion brings the best minds back home to build a better future for their children. Atlassian co-founder Scott Farquhar said that's why he and Mike Cannon-Brookes chose to build their company here:

> When we started 20 years ago we were unique, there was really no startup ecosystem in Australia. The obvious choice for where we should set up our base was Silicon Valley, but we chose to stay in Sydney because it's the best city in the world and we want it to stay that way. I know the US very well and I know Australia very well. I think we've got it better here. If we want to ensure our kids have the quality of life we do, and make sure it's a prosperous nation, we need to be creating technology jobs here in Australia.[78]

I am also confident we can unleash a new era in innovation in Australia because we have been quietly doing it, without drawing attention or complaint, for many, many years. So often, I think governments get confused by what industry wants. In Silicon Valley, there was a famous meeting when newly elected President Barack Obama came out to the Valley and gathered the area's 50 most influential leaders led by John Chambers running Cisco, Craig Barrett the Chair of Intel, and Carly Fiorina formerly of Hewlett Packard. I got to be a fly on the wall in that meeting and listen to the conversation. What stunned President Obama was Silicon Valley didn't want anything from him – the companies there just wanted to be left alone. It really was that simple. It's hard if you're a president of a country and you've got a laundry list – I remember President Obama saying, 'This is what every state's list of wants is' – to understand that all Silicon Valley wanted was to be left alone: 'Please don't try to regulate us. Don't try to legislate us. Just let us do what we do and drive wealth for the country.'

Australia has been doing amazing things with few resources, and so I'm excited by what's possible with just a few more targeted resources. Both Blackbird co-founder Niki Scevak and Spreets founder and CEO-turned-investor Dean McEvoy use plant analogies to characterise the way Australians have been sneaking across the Valley of Death, undetected. Niki said:

> I came across a great parenting metaphor a while ago of dandelions versus orchids. Orchids are a fragile plant and need lots of love and care, water it a couple of times a day, and don't let anything bad happen to it. Don't let your kid make any mistakes and don't let them graze their knee. Dandelions don't get watered, are exposed to the elements, and for some reason still survive and build resilience.
>
> In the world of tech and science, you can be an orchid who wants government sunshine, people to love them, and all the great programs, or you can be a dandelion who doesn't complain and just gets on with it. Dandelions don't care, let the fragile orchids talk about how the world isn't perfect and that they're not getting enough attention.
>
> I see Atlassian as a classic dandelion – with no money and no love, they created one of Australia's biggest companies. When they started the company, they couldn't care what the tax rate was, couldn't care whether the government supports them or not, couldn't care if they're in the media or not. For 10-12 years into Atlassian's journey, no one knew what it did or who they were, or even if they were successful.
>
> All of the successful people I meet don't care about all the other stuff. There's 'in the arena' stuff and then there's 'peanut gallery' stuff, and this is peanut gallery stuff to them. They're on the field, they're motivated by achieving the mission that they set out to do, they're internally scoring themselves to make sure they get better every day and they're their own harshest critics. Orchids are not their own harshest critic, they blame anyone but themselves for their failures, so to me, it's dangerous to cater to the orchids versus supporting the dandelions.[79]

Mike Cannon-Brookes agrees, characterising his 'dandelion' journey:

> What's really important about the Atlassian business journey has been patience. Patience for revenue to come later, but with a more solid model behind it, if you like. Part of that was we're very patient people. We're very long-term thinking. We've always had these super long-term goals for the business. The second part of that is Australia. We're a very resourceful sort of survivor nation.[80]

Dean McEvoy, founder and CEO of Spreets, also agrees, saying 'growing a startup in Australia is like trying to grow a plant in a dark cupboard':

It's really, really hard and often, most plants die. But, for those that evolve to survive in that environment, when they have some resources sprinkled on them, when they get taken out into the sunshine, they grow faster and they're more resilient and they're better off. Look at the Atlassians and the like, they scrounge together to work out how to build something without the capital resources and then you throw some resources on them and look at them fly. I think that culture of making more go further is instilled even now. You don't hear about the $20 million first round craziness that happens in the US, and I think that bloatedness gives you a different approach. [US startups] don't hustle as hard, they over-hire, they spend money on stuff that doesn't matter. Whereas I think [Australian startups] have a bit more of a focus on ROI [return on investment], which makes for a better venture capital industry and makes for better startups, and I think some US investors are starting to see that. They're like, wow, these companies actually have revenue and know how to build stuff.[81]

Matt Barrie, CEO of Freelancer.com and Escrow.com, said the constrained startup environment in Australia has made our tech companies world-class:

The really good Australian tech startups are phenomenal, because you've got to make do on the smell of an oily rag. You've got to build your businesses without funding – although that's changing – you're a market that's a long way away, it's small, you're Australian, it's hard to get attention in the global marketplace. Especially in the US. So the best we have are very, very good.[82]

Catherine Livingstone, former Cochlear CEO and former CSIRO Chair, said this ability to do impressive things with few resources has historically been part of Australian ingenuity bred from a childhood living on the land. This resonated with me because I remember spending my school holidays with my cousins in the country, and always having to carry baling wire and pliers in our pockets in case we came across something broken. In fact, when solar pumps came in, a lot of people, including my cousins, didn't want to adopt them because they couldn't fix them with a pair of pliers if something went wrong. Catherine said:

Many of our successful innovators actually come from country Australia, where there's a strong culture of, 'There's a problem, I've got to solve it, I've got to solve it now, what do I have, let's do it.' This is an appalling generalisation, but if you look at the Europeans, you often get the attitude of, 'Don't tell me what to do.' If you go to the US, where it's more hierarchical, it's, 'Tell me what to do and I'll do it.' If you go to Australia, it's more, 'I know how to solve it. Don't worry, it's okay, I've got it.' So it's not, 'Don't talk to me'; it's not, 'Yes tell me'; it's, 'Don't tell me it can't be done.' If you look again at many solutions where people say, 'It's not possible' or 'That won't work', we have a 'Don't tell me it can't be done' attitude.[83]

That attitude is a fortunate trait for a nation who will have to collaborate better than other countries because we don't have as much money or as many people to throw at the problem. That should make us more innovative, give us a more differentiated outcome, and it absolutely will make us focus. We'll use a national 'market vision' (discussed in Chapter 5) to pick a small number of things that are uniquely important for Australia to do and that will serve in our favour.

It would be easy to look at the testimonials of people who love living and working in Australia and extrapolate a bright future – but I can go one step further. There are a number of founders and startups at CSIRO's collaboration hub at Lindfield in Sydney who have sought out Australia as their innovation hub of choice. These entrepreneurs raised money overseas, completely bypassing Australia, but preferred to be in Australia – in fact, quite a number actually relocated from elsewhere in the world. Their investors are wealthy people like the Jacobs family who founded Qualcomm, or internet investor Yuri Milner who invested $100 million over 10 years to support the search for extraterrestrial intelligence, to which CSIRO's dish, Murriyang, at Parkes, New South Wales, is contributing. They are seasoned entrepreneurs who like Australia's access to talent, our RDTI and our quality of life. CSIRO only found them because they're doing deep, deep-tech, and so could really benefit from access to our facilities and researchers. Their innovation spans semiconductors, robots, genetics, medical technologies (medtech) and quantum, to name a few. Their customers and investors aren't here, and yet it isn't stopping them from being here – I think this is a sign of the future. We have a number of barriers to overcome, as discussed in this chapter, but this book is full of possible solutions to strengthen our unique and quintessentially Australian system so it can unleash the full potential of our uniquely Australian talents and advantages, growing a better future for all of us.

2
Venturing solutions

To shepherd an idea from a lab bench to real-world impact takes experience, networks, industry knowledge, customer insight and business skills – and because most science-driven ideas require prototypes, experimentation and manufacturing, it also takes no small amount of financial investment. Venture capital (VC) can help bridge the so-called 'Valley of Death' between benchtop and buyer and is one of the forces driving global innovation powerhouses like Silicon Valley and Israel. Australia's first VC funds date back to the 1970s with investments in established industries like transport and retail,[84] but it would be many decades before a fund would emerge with an appetite for risk big enough to start investing in deep-tech innovations and the reinvention or creation of new industries. Today, Australia can take inspiration from international VC playbooks as our still-nascent venture system creates an opportunity to build a unique model of venture that meets our nation's needs, while sidestepping the pitfalls of other nations. In this chapter, I discuss my experience of three 'waves' of VC in Australia, and what I think our next wave should look like.

First wave (1970s to mid-2000s)
Bankers and advisors

Bill Ferris is considered by many to be the 'grandfather' of Australian venture. Bill studied the emerging fields of venture capital and private equity while completing his MBA at Harvard Business School in the 1960s and began applying his learnings working in finance at McKinsey in the US. In 1970 he founded Australia's first VC fund, the International Venture Corporation (IVC), which inspired the creation of many other VC funds in Australia. IVC invested in industries like road surfacing and concrete, consumer finance and leasing and property services, just to name a few.[85] The 1987 stock market crash ended many funds, but soon after, Bill and his investment partner Joseph Skrzynski started a new fund called Australian Mezzanine Investments Ltd (AMIL). By the late 1990s, AMIL had co-invested with the Australian government and the Walden International Investment Group of San Francisco to form the AMWIN innovation fund, which would go on to make Australia's first landmark investment in a tech startup, LookSmart.

I was completely unaware of Australia's slowly growing VC landscape while studying and beginning my own career as a founder in the US in the 1990s, although I did know that Australian scientists with ideas for businesses came to the US to commercialise their research because they couldn't – or didn't know how to – at home. Around the year 2000, as Aussie patriotism was riding high among expats in the US in the lead-up to the Sydney 2000 Olympic Games, I met fellow Aussie expat David Cannington, who was running the ANZA (Australia New Zealand) Technology Network. ANZA had been started by Belinda Cooney and Dan Phillips of Macquarie Group while they were exploring the US VC market, with Guy Manson as their Chair. At one of David's events, I met Evan Thornley and Tracey Ellery, the Melbourne couple behind Australia's record-setting NASDAQ initial public offering (IPO), the first backed by Australian venture: an internet company called LookSmart. Evan and Tracey, then based in New York, had discovered the idea for LookSmart, an internet search system, buried within the corporate giant Reader's Digest and thought it could be a winning standalone business. In 1995, they took their idea to Bill Ferris, and in 1997, Ferris's AMWIN fund, together with a number of other investors, invested in its first round. By August 1999, LookSmart was the first business backed by Australian venture to debut on the NASDAQ and set an Australian record for the largest IPO, valued at US$1.7 billion, or A$2.5 billion.

After meeting Evan and Tracey in 2000, I was impressed with the LookSmart story and excited to consider what this might mean for the growing maturity of the VC system back home if a tech startup like that could raise funds and deliver this kind of IPO, largely from Australia. A few months later, David Cannington convinced a few of us to fly back to Australia with him for what he was calling the 'US expats tour'. We met with the federal and state governments to talk about our experiences, and I started to understand the depth of the science commercialisation problem for Australia – what we now call our innovation dilemma. The aspiring Aussie entrepreneurs I met didn't know what to expect in the Valley, so they made assumptions based on their interactions with Australian venture capitalists, who, at the time, had mostly been bankers and who were more interested in the financials of the proposal than the technology behind them. For every scientist and would-be deep-tech founder trying to get a great idea off the ground, there was an advisor they'd hired to help who would be calling the shots and pulling the tent down around them. They would prepare for meetings with venture capitalists the way they would prepare for a home loan application: by focusing on a flash set of numbers to make their case. There were a lot of consultant middlemen in Australia who would say they knew 'the venture game' and they'd offer to navigate the entrepreneur through. A lot of them talked like game show hosts and promoters, and it was very hard to disabuse the scientists of the notion that's what they needed to be. We had these amazing, genius scientists who were phenomenal – and then we'd have these ringmasters and showmen telling them 'how to handle the venture capitalists'. They had the razzle dazzle of the publicity and media agent Harry M Miller or notorious businessman – and eventual fugitive – Christopher Skase. Even worse, some Australian startups got so tied up in lawyers that when they came to the Valley, their companies had complex structures with IP owned by a company in another country, no stock, shares held by another company, a nested structure for tax reasons, and other features their lawyers advised. The complexity meant US venture capitalists didn't know how to invest in them because they didn't really know what they were investing in.

Coming from the Valley, I knew they were a very different kind of VC to the Australian model. Valley venture capitalists saw themselves first and foremost as coaches or partners to their founders. If you're a fair dinkum tech or science entrepreneur, you're better off being a little bit clumsy and awkward but deeply knowledgeable about your science, because there's an authenticity to that. Most scientists I've met have that courageous authenticity, a wonderful attribute where they'll say things that other people might be afraid to say, or won't say, but they're just being themselves, just being honest; they're sharing with you what they really think. Those are the kinds of founders who secured investment and built the Valley, and those are the founders we need to grow and support to become our next generation of CEOs.

Good Silicon Valley venture capitalists were more interested in asking, 'Who am I really working with here? Can I trust you? Will you listen to me when I give you advice?' When I was a venture capitalist, there were three questions I'd ask before I'd take a meeting with a founder so I knew I wasn't wasting my time and that I'd actually be able to work with them. The first question was: why do you want to do this? It is my missionary versus mercenary test, discussed more in Chapter 3. The second was designed to see whether they'd listen and were interested when I'd mention that I did my PhD in physics and lasers. The third question was to understand if they had ever failed and what had happened.

Generally, Australian entrepreneurs don't want to admit failure. If they do tell you the story of their failure, they'll always be the star of the story, and it won't really have been a failure. It's very similar to what I've found when asking a male CEO to talk about their diversity learnings – often the hero, rarely the learner. The entrepreneurs I want to work with are the ones who have the courage to tell you about their mistakes. You'd be amazed at how many investors prefer to back someone who has tasted the bitter fruits of failure. In failing, you learn what not to do. Get your skin in the game and there is no failure – you have opened your mind to growth and yourself to reinvention. Venture capitalists invest in people first and ideas second.

The Valley back then focused on deep-tech science plays, had a severe allergic reaction to those 'showmen advisors', and it took a generation to figure that out. It is amazing how badly we listen. Hollywood has done such a job on creating a mystique for the 'shark tank' idea of venture, ruthlessness and the rest of it. The only thing I've seen from Hollywood that came close to true venture was the movie *Extraordinary Measures*, which is so true to life because it was based on the true story told in Geeta Anand's *The Cure*. Brendan Fraser plays John Crowley, an advertising executive whose children are dying of a rare disease, so he contacts Harrison Ford's character Dr Bob Stonehill, a scientist whose research into the disease shows some promise. They decide to pitch to a Valley venture capitalist for investment to fast-track the research, and Fraser's character does what every Aussie entrepreneur does: shows the spreadsheets and the analysis and the 10-year discounted cash flow and the business plan and it's all perfect. The venture capitalist – who is also a scientist – says to John: 'If you were still in business school, I'd give you an A for all those lovely charts and graphs. But school's out. Most of us here are scientists. We need to see the science. Bob, make us believe.'[86] I've never seen Hollywood get it so right about venture. The really good venture capitalists, the ones scientists should want to work with in the Valley, start with those questions. Tell me about why you want to start this company and what's driving you. When you asked Aussie entrepreneurs those questions back then, they looked at you like you were crazy. Why do you want to know that? That's a stupid question.

I realised that there were many Australian scientists who could be entrepreneurs, because they were a lot like me, but they were never going to succeed in Australia with Australian venture capitalists at that time. I wasn't the kind of CEO that Australian investors would jump at supporting – I wasn't a McKinsey graduate or even an MBA graduate, but I had already co-founded multiple companies in the US. Something wasn't connecting properly in Australia between scientists and investors. I began to understand why Evan and Tracey from LookSmart weren't scientists like me, despite running a tech company. The LookSmart founders had the kind of business pedigree that mirrored that of Australian investors. They were both graduates of Melbourne University: Evan had gone on to be a leading consultant in telecommunications, consumer goods and energy at McKinsey, and Tracey had a career in computer retailing, market research and publications.

The more I got to understand the state of Australia's narrow, banker-driven VC system, the more I began to see this was limiting the system's potential to support deep-tech companies. For example, it was an absolute non-starter in the industry that a deep-tech founder could be the CEO of their startup. As soon as they invested, venture capitalists were thinking about replacing the founder from day one – it was the first step in the VC playbook. Do the investment, get rid of the founder, bring in a safe pair of hands as CEO. The notion that 'we're professional investors, we know better' went on for decades and yet we kept wondering, why are we failing? That was probably one of the classic failure points of Australian venture back then – and it still happens today. I talk more in Chapters 3 and 4 about why scientists can and should be the CEOs of deep-tech startups.

David Cannington's tours ignited my passion to do something to change the system, so I started dedicating a lot more time to David and his group. We did more of these tours around Australia, we had more Aussie entrepreneurs come to the Valley, and I got involved in various startup functions and events to try to stimulate more US VC investing in Australian entrepreneurs, filling the gap created by the immature VC ecosystem at home.

The Australian Securities Exchange (ASX)

A few years later, an opportunity shifted my focus from being a US-based expat trying to support Aussie entrepreneurs to becoming an Australian-based founder, right in the thick of the problems. In 2004, my PhD supervisor at Macquarie University in Sydney, Professor Jim Piper, convinced me to come home for a while and try my hand at being a professor with him. The Australian government made me a Federation Fellow, a prestigious position designed to lure Aussie expats to come back, and so, for about a year, I was Professor Marshall at Macquarie University. Now, I've never, ever, used that title because it would forever brand me an academic to Australian business and simultaneously enrage the real professors in academia who truly earned their title – it's very hard to be both in Australia. While I was oscillating back and forth between Silicon Valley and Australia, my old friend Xiaofan (Simon) Cao proposed building on the concept of the optical chip I'd developed in my company Lightbit, so a few of us founded a company called Arasor to develop the idea. I divided my time between Arasor and being Professor Marshall, and by the end of that year, Simon became Chair of Arasor and persuaded me to become managing director and run it. I couldn't last a year as a professor before I ended up running a startup again. I still laugh to think of myself as a professor because I have never really fitted that mould. That time operating in the Australian system gave me great lessons, including acquiring some excellent IP from the University of Sydney and leveraging a small semiconductor fabrication plant (or fab) in the university's 'Bandwidth Foundry', to build up Arasor in Australia, China and the US. That first acquisition was another lesson in culture difference. The venture capitalists around the table who had decision rights in the Foundry actually hadn't invested anything in it, but they did control it through another investment they made years before. They weren't going to support it in the future – and yet they still wanted to get paid for us taking it over. This was a surprise to me because in the Valley, venture capitalists just walk away when something fails. They know life is too short to waste time on trying to claw back returns from a failed investment and that you're better off spending your energy on making more suitable future investments. In a 'finite pie' ecosystem you squabble for every penny; in Silicon Valley you just make more.

Arasor was a tale of two cities, or rather two technologies. It had amazing optical technology and would go on to supply unique communications gear for the 2008 Beijing Olympics. It had a wireless system that could stream high-definition video at carrier-grade quality, enabled by a unique optical chip technology. We take this ability for granted today, but imagine having it almost 20 years ago, and imagine what Netflix would have done with it back then!

Arasor had great tech, but we were competing against much larger incumbents and it was hard to get a big telco to buy into the vision of a small private company. Simon was convinced we needed to IPO to have the credibility and governance that large telcos needed to see, but the new Sarbanes-Oxley (SOX) requirements for listing on the NASDAQ, introduced earlier in 2002, set the bar too high for revenue for us (later in this chapter I discuss why it was a different story for Iridex in 1996). I got to know a few of the underwriters in the Australian banking community and they agreed we should take Arasor public, so joined our brains trust as we started planning.

The nirvana for US venture capitalists was always to take their companies public on NASDAQ, as LookSmart had done. We began to wonder: why shouldn't we be able to do it here in our own market, where we have our own capital? When an Australian company listed on the NASDAQ, there was an expectation that it would relocate to the US to satisfy the US analysts' need for access and to secure US investor interest and support.[87] My theory was, if that ecosystem was created in Australia instead, and we could make it as appealing to list on the ASX as the NASDAQ, it would start to create an ecosystem here where companies wouldn't need to leave and would be able to grow here. We thought it would be worth trying to take the first Silicon Valley tech company to IPO in Australia, and in October 2006, we did. It wasn't easy. But like so many things in Australia, it needed to be done overseas first before it was recognised back home, and if it could be done it might unlock more opportunity for domestic startups to follow the same path. The Australian market was suspicious because it had never been done before and couldn't understand why a Silicon Valley company would IPO on the ASX. There was no tech analyst in the Australian market back then who could follow and report on a tech company. It would have been so much easier if we were mining the chip rather than inventing and building it ourselves – we would have had dozens of analysts who understood the resources sector.

A full decade earlier, Catherine Livingstone found the same challenge when taking Cochlear public at the end of 1995:

> There was no biotech, biomedical, there was no sector that was acknowledged as such. There was no analyst coverage of it. The joint lead managers who were doing the IPO had to encourage some sort of coverage. I remember they sent along Brett Clegg, [who would go on to be Managing Director of News Corp Australia] who was very early in his career at that point. They told him, 'Go forth and find out about this sector', and to his credit, he did a really creditable job at the time with nothing to call on or rely on. So it was difficult and it was also difficult going through the US. I remember on the road show, no one wanted to talk to us, you lost significant points for being Australian. 'Where's Australia? What do you do? What's a cochlear implant? Don't know, not interested.'[88]

Arasor was one of the top three IPOs in Australia that year. It got investors even though there wasn't a market maker (analyst) in deep-tech. We found a lot of willingness from the big funds to dedicate people to learn about our market, to investigate our technology and actually make a market. Three analysts latched onto the Arasor story and continued to follow and report on the stock, and they went on to explore other tech stocks; this was critical to growing the tech market.

Arasor did something no one had ever done before. While I was Managing Director of Arasor, we grew our revenue from $5 million to $35 million. The year after our IPO, when I left the company, it delivered just under $120 million in revenue. I left the following year, and it delivered three times more revenue than the prior year and its share price increased 270 per cent. We built strong relationships across the sector, like with Gary Wiseman who was at Intel's optical communications group and who now helps Aussie startups like Baraja (discussed in Chapter 1) in Silicon Valley.

After we completed the Arasor IPO, the investors, including Macquarie Bank, which was the cornerstone investor in our roadshow, started talking about forming a venture fund to fix many of the things we'd experienced as founders of Arasor and saw other startups struggling with. We were frustrated that founders basically had to get on their hands and knees to get investment from VCs. They had faced 70-page term sheets – the contract signed by founders to access investment – which were full of hooks and buzzsaws and bear traps, basically asking you to mortgage your home, secure all of your assets, donate a kidney and offer up your first-born child. Blackbird co-founder Niki Scevak posits that the elaborate term sheets were never about risk minimisation, and were actually just evidence of an unwillingness to embrace the inherent and necessary risk of venture:

> In life, when someone has no knowledge of something, they tend to ask more questions. When we [Blackbird] raise money from investors, the ones that you know aren't going to invest will send you a 100-page questionnaire. It's a reaction to a deep insecurity. When you don't know anything, ask 100 questions. The 70-page term sheet is the equivalent of not knowing anything, and as ex-bankers and ex-lawyers, it's their outlet for the 100 dumb questions. It's a signal of their insecurity that they don't know anything and so they're channelling all their energy into something that ultimately doesn't matter.
>
> The 70-page term sheet is also a philosophical difference of downside minimisation or downside protection versus [knowing that] in venture capital, you don't achieve success by getting 70 cents on the dollar or 90 cents or 10 cents. It doesn't matter. You may as well lose the whole dollar and spend no time on it because then you have time to spend investing in the ones that are a 1000 X [will make you 1000 times return on your investment].
>
> For a 1000 X outcome, conceivably you wouldn't even need a term sheet, you would just need to have the equity ownership in common shares because ultimately that's the upside. Now obviously, there's the reality of the world, and there's the market and so on, but on some philosophical level, you have to decide, do you believe that you'll make

money with successful investments or do you believe that you'll make money on minimising the unsuccessful investments? The 70-page term sheet is squarely in the thinking that minimising unsuccessful investments is what will make them successful.

That approach is true in private equity. In private equity, it matters that you don't lose very often and you at least recover your capital. For some of them, you might make 2 times your money or 3 times your money. But it really matters that you don't lose your money. It's a logical and rational thing to do in private equity, but in venture, it's irrational.[89]

Everything was about managing possible downsides; many venture capitalists had still not recovered from the shock of the 1987 stock market collapse. So we set out to create a new wave fund that would be lighter on the paperwork and faster to seize opportunities – more Silicon Valley-like because it would operate on both sides of the Pacific, in Australia and the Valley, and be run by founders for founders.

Second wave (2007 to 2012)
Bridging investment ecosystems

In 2007, the year following the Arasor IPO, I stepped away from the company, as we'd told investors I would. We started building a team around the new fund that would become Southern Cross Ventures. An American living in Australia, Bill Bartee, joined us from the investment arm of Macquarie Bank, where he'd also invested in LookSmart and Seek. I'd met Bill years before through David Cannington and discovered the small US expat cohort inside the Aussie startup community – the inverse of my experience.

Bob Christiansen and John Scull left the Australian VC firm Allen and Buckeridge to kick off Southern Cross. Bob was a successful entrepreneur on the US east coast and returned to Australia after about three decades to join Allen and Buckeridge, which was one of the largest Australian VC funds at the time after Bill Ferris's AMIL fund. Bob is a software expert and is the same 'entrepreneur' personality type as me – so much so that, in addition to annoying each other greatly, we both went through a similar personality transformation that entrepreneurs can go through as they get older, transforming from entrepreneur to CEO type. I talk more about these personality types in Chapter 3. Bob's CEO personality used to really annoy me when I was still entrepreneurial, in the same way my CEO personality can annoy entrepreneurial spirits today.

John Scull was more the CEO type than Bob and me. He worked at Apple and created what later became graphics and web development company MacroMedia, which was then acquired by Adobe. John's job at Apple was managing Steve Jobs's creativity, picking which ideas were genius and trying to get Steve and Steve's successor as CEO, John Sculley, to focus just on them. John Scull (not to be confused with Apple CEO John Sculley) likes to move quickly and, while being a market visionary, can really challenge entrepreneurs, especially Aussie entrepreneurs.

Gareth Dando joined after being CEO of Australian university-based VC fund Uniseed and is a brilliant strategist and analyst. He was also a bit of a 'scientist' personality-type, in that he could see the myriad of possibilities but it was hard to get him to lock in on a decision.

The team evolved to fit the diverse roles we needed to manage the fund and helped me understand this secret recipe of personalities for running a VC firm – a recipe Bill Bartee and I used a few years later to create the right cast of characters in Main Sequence (discussed in Chapter 6). For example, Bill and Gareth knew when to exit an investment, but Gareth and Bob had trouble knowing when to get in. Once we were in, Bob knew how to really commit to backing a winner, when often the Aussie VC approach is to stop investing and let others carry the load when the valuation gets too high. John had great 'market vision', or ability to invent an imagined future (discussed more in Chapter 5) from his days at Apple, knew that disagreement and controversy were key to making good investment decisions and could back the unpopular view (also an Apple skill). I was the entrepreneur who couldn't stay on mandate or follow the rules and annoyed everyone, but our combination worked, because we were such a diverse group in our life experiences and personality types.

Unfortunately, that diversity didn't extend to other aspects including gender, just like the rest of the VC community at the time and, to some extent, still today. Fortunately, soon Tristen Langley joined our motley crew in 2007, bringing her learnings from top-tier Valley VC firm Draper Fisher Jurvetson (DFJ). Tristen was another Aussie expat, who got her start in venture with first-wave Australian venture firm Allen and Buckeridge just as the dot com bubble was rising and the internet was beginning to take off in Australia. She compares the experience of leaving Allen and Buckeridge in Australia for DFJ in the Valley by saying, 'Allen and Buckeridge was terrific because they blew the doors open to what was possible in the VC environment in Australia, but in the next wave, I joined DFJ in Silicon Valley and that opened my eyes to everything that was possible, because Silicon Valley is the foothold into the intersection of leading-edge technology and relentless capital.'[90]

Together, Tristen and I did almost half the deals in Southern Cross. Tristen was a powerhouse of energy just waiting to be unleashed, with a brilliant mind and a Stanford MBA. She talked me into backing Quantenna and Wave, both of which became 'unicorns' (companies worth over US$1 billion). She tried to talk me out of investing in Redfern Integrated Optics (RIO); that was when I learned to regret not listening to her (all discussed later in this book). She drove me crazy for the first year, but I learned to love working with Tristen. We rarely agreed on anything, but I think that was the key. As opposites often attract, our collaboration generated amazing energy. Tristen went on to become the lead Series A investor in MyGlam/Ipsy, which is now the successful personalised cosmetic subscription service Beauty For All Industries (BFAI), another unicorn.

Southern Cross was the first fund to operate both in Silicon Valley and Sydney; this meant we had created a bridge to link Australian founders here to the money in the US. Our first fund was about $180 million – one of the biggest funds ever raised at that time – but $180 million is still only a small pot to spread around capital-hungry deep-tech startups. That is why we were a bridge to more funding in the US – we couldn't be the whole freeway. Freelancer.com and Escrow.com CEO Matt Barrie recalls the challenges of small Australian funds who weren't connected to the US venture system:

> They would put money into the startup, you'd get going, you do all the hard work, all the risky stuff early on. That first million of revenue is the hardest, you're solving complex problems, all the problems that happen with the team happen early on typically. You'd do Series A fundraising in Australia, maybe Series B, but then you'd go to the US because you needed to raise more than you could in Australia. The minute you 'flip up' to move to the US and access its funds, the first thing the American investors would say was, 'This company is under-funded and you need an American CEO and American management team.' The American CEO would come in and say, 'This company's under-capitalised, let's raise more money. Let's raise a lot more money.' That would then dilute your Australian investors, who would get tapped out pretty quickly. They did all the hard, early stage, high risk work, and the Americans swoop in with the money and start replacing the management team.[91]

CEO of Australia's Future Fund Raphael Arndt said the fund invested with larger US venture firms before eventually investing in Australian venture for exactly that reason: to avoid being diluted at the wrong time.[92]

From the start, Southern Cross set out to be different. It was the real start of my journey towards solving Australia's innovation dilemma (outlined in Chapter 1). We fought and fought and fought with our lawyers about 70-page term sheets, not to mention with our investors, who also wanted water-tight investments. As a compromise, we put our term sheets up on our website to make them easier for founders to access, although those term sheets were still longer and more complex than they needed to be. Part of being 'more Silicon Valley-like' also meant we were looking at the market and really trying to predict where things were going to go using a 'market vision' (discussed in Chapter 5). We specialised in finding promising Aussie startups that would disrupt or pre-empt those markets and then helped them tap into professional venture.

Southern Cross was a complete shift in how venture capitalists operated in Australia, from dealing with bankers and 'just the money' to being a fund with people who can really understand and help your deep-tech business grow. Because we wanted to differentiate ourselves from the rest of VC in Australia, we didn't do the 'pitch model', where founders were expected to come in and give venture capitalists an elevator pitch on why they should invest, which felt like crawling over broken glass and begging for the money. Instead, we went 'old school' Silicon Valley: after going out to cherry-pick founders whose research matched our market vision, we did a lot of 'sweat equity', where you earn your share of the company's success by rolling your sleeves up and working hard alongside them.

We also worked hard to earn trust, because Australian venture had a bad reputation among entrepreneurs for being difficult to get and lacking in support to help a company grow. The more we worked with founding teams and influenced how they grew their companies, the more we lowered adversarial barriers between the venture capitalist and the founder; this made for much more cohesive management teams and better outcomes. We spent a lot of time on that and then we'd bring them across to the Valley in cohorts to meet with our investor partners. Sometimes we'd even set them up in our office in the Valley to get the company going there.

It wasn't the purest form of the model because I knew what I really wanted in the long term was to have them raise money here, grow their employee base in Australia and contribute to Australia's economy, and I knew investment from a US venture capitalist would eventually mean it became a US company. But back then I also knew they wouldn't get the buy-in of the broader investment community in Australia, so the best place to go for the money was the Valley and Southern Cross could be the bridge they needed to get to the money and networks of the Valley.

We were making progress in shifting the way deep-tech founders thought of venture capitalists, but it took much longer to shift the way other Aussie venture capitalists thought of deep-tech founders. Lawyer and advertising executive

Jodie Fox, together with husband Michael Fox and friend Mike Knapp, turned her passion for shoes into the startup Shoes of Prey in 2010.[93] When Southern Cross invested in them in 2012, we could see that Jodie would be the natural CEO when it was time to formalise the company's structure – she knew the market in her veins. But when the investors looked at her beside her husband, he was the one who looked like a 'traditional' CEO, so he was given the position of CEO. Atlassian co-founder and co-CEO Mike Cannon-Brookes was another investor in Shoes of Prey and later he joined their board. When I first reached out to Mike about this deal he said, 'I'm already looking at it, so I guess that makes us competitors.' It didn't take much to convince him we should be collaborators instead, and it was great of him to let us do the deal together when he could have just steamrolled us instead. When you're a successful internet entrepreneur like Mike and have established yourself as a trusted advisor, there's a peril that you'll throw off the instincts of a founder who knows the market better than you do. As Mike was a newer investor, we talked about how he would need to be conscious of giving advice on the areas he knew about as a fellow internet startup, but also knowing when not to offer advice. Shoes of Prey soared before it collapsed in 2018, discussed further in Chapter 3.

Southern Cross also attracted other entrepreneurs who wanted to get into VC. We brought in successful internet entrepreneur Michelle Deaker, who worked out of our Sydney office while she figured out how to raise her own fund, which became One Ventures. Michelle had a really tough time raising that first fund during the Global Financial Crisis and it gave me a glimpse into some of the inherent biases and challenges in the Australian system. Michelle also told me terrible stories about capital raising as a female entrepreneur. In the end she became one of the few funds to transition from the second wave of VC that characterised Southern Cross and others to the third wave of VC, discussed next. Michelle was not the first female venture capitalist in Australia; before her there were also great people like Fiona Pak Poy, Brigitte Smith and Teresa Englehard to name a few, but Michelle became one of the first women to found and lead a VC firm as Managing Partner. Today there are more, like Lucinda Hankin heading up Grok Ventures, Melissa Widner at Lighter Capital and Kim Jackson at Skip Capital. Kim came to see me a few times to ideate and I really enjoyed sharing my thoughts on how funds work and how they should work – but what I loved was the way she listened, said 'That's nice, but I'm going to do it my own way' and she did. Fund managers need to be agile and flexible like entrepreneurs, constantly reinventing, which Kim did.

Navigating the Global Financial Crisis (GFC)

The real star of the Southern Cross portfolio was Quantenna, which was already close to my heart for its connection with CSIRO, where I'd been a summer intern in the 1980s, as well as for its similarities with Arasor, which solved the bandwidth problem on the carrier side while Quantenna would solve it on the customer side. Southern Cross helped build Quantenna by bringing together people with expertise in wi-fi from what was then National Information and Communications Technology Australia (NICTA), as well as from CSIRO where fast wi-fi had been invented 20 years earlier. We got them talking to like-minded folks at Stanford to develop the next generation of wi-fi that could beam through stone walls, completely eliminating the need for cable in a home or even in a large enterprise. Leading Valley venture capitalist Sequoia invested in it even before we did, because we wouldn't invest until there was a clear Australian opportunity and it took a while for the local engineers to make the leap on board – though we did invest a bit later and on every subsequent round. It would go on to IPO in October 2016 on the NASDAQ, arguably the first deep-tech Australian VC-backed company to do so – if you consider LookSmart to be software, not science. I talk more about the difference between software and science startups in Chapter 3.

Southern Cross had been running for about a year when we first started to hear rumblings of what would become the GFC in 2008. Southern Cross lost about half of its portfolio, but the companies that survived were the science-driven ones that had set out to do impossible things and their uniqueness saved them. The excitement around what Quantenna was going to do meant that by the time the GFC hit, it had five funds around the table with a lot of capital and a lot of belief, so we all chipped money in to carry the company through the GFC.

In a downturn, you need something to really differentiate your startup. Having a technology that's extremely compelling will push through even the most negative of markets. You only need two parties who both want the deal in order to make the price higher than it would be if there is only one; if you're a unique company, you might have 10 parties all trying to invest in you and suddenly you're back to competitive market demand conditions. That's absolutely what we found in the GFC, and not just in my fund – I think all funds learned that early stage, high-risk technology, and products and markets that haven't been invented yet, are upsides in a downturn. That goes especially if the company isn't in a growth phase and needing significant capital yet, but is in the sweet spot of having proven the technology, having proven the market need, and just needing enough investment to carry through to its next company milestone.

The GFC really broke up the party on VC. For the past decade it had been a bit like a drunken gathering of investors piling in to capitalise on the big wins of the early 2000s and now they were walking away from venture, nursing their hangovers and swearing never to invest again. In Australia, Macquarie and many of the big super funds walked away and the 'fund of funds' model – where investors can invest in a VC firm rather than directly become VC investors – that had been supporting VC more or less collapsed after the surge in 2006. It was about 7 years before super funds put a toe in the water again. This pattern of jumping on the bandwagon after the boom has happened is part of our 'innovation follower' mindset. After that, the mindset of super funds in Australia was such that they weren't even investing in US venture, let alone Australian venture. Hostplus became a first mover when it dipped a toe in the water in 2011 by investing in US venture firms, then dipped a foot as it cautiously invested in Blackbird in 2014, and then a whole leg more recently as it's become a primary investor itself in areas like fusion (discussed in Chapter 6).

I'd met with the Chair of the Australian government's Future Fund, former Treasurer Peter Costello, a few times. He said the fund would never get involved in venture, it was simply too risky, and the government was accountable for public servants' retirements. Ironically, Peter said the Future Fund wasn't very well named because many people had suggested using it to change the future by supporting Aussie venture or science – I think he meant that it isn't really about creating the future but rather protecting our future. I could never persuade him to invest in anything before the GFC. But once the GFC hit, US venture capitalists couldn't raise any money and Peter and the Future Fund bought into these funds at the worst time for the Valley and the best time for them. They've done really well in US venture as a result of that, with their venture exposure now worth about $5 billion.[94] I remember being at an AVCAL (now Australian Investment Council) event in the middle of the GFC and Australian investors were swearing off venture for life, but Future Fund CEO, Raphael Arndt, reflects now that the Future Fund realised that made it the best time to get in – although they got into US Venture initially, before eventually branching into Australian venture:

> At that point we held a lot of cash because the board under David Murray had made a very bold call to stop investing the cash box. Everyone was telling us, 'It's too late, you've missed the boat, there are no good assets left.' The culture and DNA of the Future Fund has always

been strategic, and I would even say intellectual, in taking a deeply researched view of how to succeed in the long run as an investor and didn't get too caught up in noise and short-term market issues. The research told us that around 86 per cent of venture capital managers lost money in the long run, but that, unlike every other asset class, in venture that performance was sticky – that is, when a venture firm is successful, it then attracts high quality entrepreneurs who want to go to the successful firm to capitalise on their insight, and they continue to be successful, and then it's very hard for new firms to break into that 'winner's circle' in venture. In other words, the message for us was, if you can't get in with the best managers, don't bother at all. That was the message.

The GFC presented a confluence of events. We had plenty of capital and suddenly, partly because of the tech wreck where venture managers didn't do that well, there was a withdrawal of US capital from venture and other funds, like endowments, had quite significant liquidity problems and in fact were ringing up their managers and saying, 'Don't call us, we're not going to fund you.' So US venture managers, who usually would be hard to get into, especially for a new fund on the other side of the world, suddenly thought, 'Wouldn't it be good to diversify to someone from some other part of the world who might have a different motivation.' The Future Fund thinks about the intellectual reason for doing something and then we back ourselves, so it didn't make any sense in what was then a $60 or $70 billion fund to put in $100 million or $200 million, so our first cheque was about half a billion dollars, and we decided to build from that.

We did get interested questions from government and others about whether we were or weren't backing the local scene. As a long-term investor we certainly think it worth investing in improving, or playing a role in improving, the ecosystem over time, that's in everyone's best interests. So we asked our US-based managers, Horsley Bridge and Greenspring, to spend some time in Australia looking at investment opportunities and, in fact, we explicitly carved off part of the Greenspring portfolio for Australian venture capital managers. They started to come out to Australia based on that, and I don't think they were planning on coming to Australia prior to that.

We also asked them to convene some workshops for Australian managers who were interested in hearing from some of the best in the world about what they looked for. What sort of traits were positive and what sort of traits were a worry – and we thought that could help develop the local industry and find opportunities for ourselves in the process. We ended up backing Blackbird through that arrangement, on the recommendation of Greenspring.[95]

By the time the Future Fund was investing in the local ecosystem, a new wave of venture had formed in Australia, one that went further towards putting scientist–founders at the heart of Australian venture.

Third wave (2012 to present)
For founders, by founders

In the early 2010s, whenever I was visiting Australia from the US, I'd spend a day with Startmate, a network that connected established entrepreneurs with young Australian founders and companies in need of mentoring and advice as they built their businesses. It had been started by Niki Scevak, who had moved from his hometown of Newcastle, New South Wales, to live in New York for a while and founded two software companies there, before returning to Australia. Like me, he'd made the move from deep-tech founder to deep-tech investor. Startmate brought in a wide range of mentors, from second-wave investors like Bill Bartee and me through to newer investors like Mike Cannon-Brookes and Scott Farquhar, co-CEOs of Atlassian.

As we got to know more of these younger founders, it became obvious that there was critical mass in digital capability for a new digital-focused fund to support them – but they'd need a new wave of VC with founder-friendly terms and fast decisions, because the digital world was moving fast. They were all inexperienced and we knew no one in the traditional Australian venture community would get behind them because they were fresh out of university and so had no financial or business track records yet. In fact, one of them was still at university, juggling classes while he tried to build a business around his invention: sunglasses with a chip in them so you couldn't lose them.

Niki had come to the Valley a few times to get familiar with our networks. He and I spent time talking about the possibility of a new fund with Rick Baker, who was working for MLC private equity in Australia and had come to the Valley to find a way for MLC to get into US venture. Perhaps not surprisingly, Rick couldn't convince MLC – the leadership thought it was too risky. Rick and Niki were a natural pair to begin the new fund, which would go on to become Blackbird Ventures. From the get-go, Blackbird was designed to be a fund for founders, by founders. Its partners would all be science-driven founders – whether hardware, software or digital – and they would know how to nurture and grow entrepreneurs like the talent we were meeting in Startmate. Niki remembers the 'paradox' of the Australian venture and startup scene at the time, but it created a perfect opportunity for Blackbird:

> There was a paradox of not many investors focusing on a global software company created from Australia, but on the other side of the coin, you had all of these already successful companies, obviously Atlassian being the best example, but there were probably 20 more, like Campaign Monitor, Aconex, RetailMeNot, WiseTech, Xero, the list goes on, of

companies that were already successful and showing promise. That paradox of no investors concentrating on a group of companies, and those companies succeeding despite no capital, was the opportunity for Startmate and for Blackbird soon after.

The magic of Silicon Valley is when someone creates a company, they invest and help the next generation, that is the true magical circle of life that ultimately makes an ecosystem bigger over time. So it was also very clear that if you got this activity from no money and no help, then if you could bring together the founder community and some capital, then this group of successful companies could be a group of successful companies times 10 in the next generation.[96]

Blackbird started within Southern Cross initially, because raising investment was still difficult so soon after the GFC. We went out to our own networks and each raised about $10 million. Niki and Rick raised mostly from Australia and the current generation of entrepreneurs, while I raised from the prior two generations of both US and Australian entrepreneurs who wanted to give back. Mike Cannon-Brookes was the first and biggest investor in Blackbird and Dr Peter Farrell, founder of medical device company ResMed, was the next one in. I loved that, straight off the mark, we had an entrepreneur from the era of the first wave of Australian venture represented in Peter, although he founded ResMed in 1989 without any venture investment; the second wave of Australian venture represented by my background with Southern Cross; and the third wave emerging with Mike, who was starting to leverage the success of Atlassian into investing in other founders. I thought that if we have three waves of entrepreneurs in one fund, there would be magic here that we can use to nurture the ecosystem. I talk about the importance of this 'pay it forward' mentality in Chapter 3. That was the common thread across all three waves – although they had different ways of going about it, they each were really passionate about showing that Australian entrepreneurs were just as good as Silicon Valley or Israeli entrepreneurs.

Despite this early interest, we reached a point where we were really struggling to raise the money for Blackbird and it didn't look like it was going to be successful. A young investor came to us from Victoria representing about $40 million lined up through very high net worth Victorian families, who wanted to invest with us. About half our partners wanted to bring him in just so we could get the fund underway. It was one of those classic moments where you've really got to question your conviction and your principles. It has happened in every startup that I've ever been involved in where you seem to be inevitably faced with both an easy choice and a hard choice. It would have been so easy just to take the money and tick! we'd be ready to go with the fund. But I believe that if we'd done that, it wouldn't have been anywhere near as successful as Blackbird went on to be, because we stuck to our principles. While there wasn't anything wrong with his intentions or the investors he was representing, they weren't founders and I believe that the secret to this fund was the founder network. It was the first venture fund that was created by founders, not by bankers or family offices or professional investors. That was Blackbird's moment of crisis and it reaffirmed my belief that every startup has a moment where it's either the thing that makes you fail, or you rise above it and you turn failure or near-death experiences into success (discussed further in Chapter 4).

By March 2013 we were ready to officially launch Blackbird outside of Southern Cross, which Bill Bartee left to join Blackbird. We started with $29.5 million in the fund from 90 names in the Australian entrepreneur community who had contributed. By building this network of 90 entrepreneurs, we knew almost everybody in the Australian innovation system and it meant each startup was only one person away from whoever they needed – be it a customer, employee or whoever. The collegiate power of that group was amazing, similar to the dynamic of the Aussie startup network David Cannington nurtured in the US.

We'd tried to get government involved through its Innovation Investment Fund (IIF) program, but once again we were deemed too risky. To be fair to the government, it had the same mindset as the venture capital community at that time: digital and internet startups were too risky, because the first few companies Blackbird invested in had failed. That's partly because we only invested in risky, early stage companies. We didn't have anyone to impress because the 90 entrepreneur–founders knew that world – they'd all lived it. With no government investment and only founders involved, we didn't have to play the game of balancing investors' need for early returns with our own curiosity and desire to push boundaries.

Leading by failing

We were free to fail as investors and we learned a lot along the way. There was a wonderful company called Ninja Blocks with a really passionate founder and the brilliant idea that you could have little bits of hardware that would fit together so you could make almost anything. Imagine you wanted to build a home automation system. You just get a few of these Ninja Blocks together and you've got it. It was like digital Meccano. Niki loved this founder and it was the first time we broke our mandate of focusing on digital, but you've got to back the founders you love. Although it failed, it failed for all the right reasons. It failed because the team were trying to do something hard. It failed because, even though the team did everything they could to make all the right decisions based on what they knew at the time, no one could predict that particular market. Failing for the wrong reasons would have meant they'd been arrogant, thinking they knew everything and ignoring all the market signals. These founders really learnt from that failure – as did we. If we were a traditional fund, the lesson would have been enforced: stay on mandate. But if we'd done that, the fund wouldn't have been as successful – if you aren't failing more than succeeding, then you aren't trying hard enough!

I've always believed in the importance of thesis-driven investment, where you invest in companies that align with your market vision (discussed in Chapter 5). One day Niki suggested we consider an autonomous car idea called Zoox – this was not digital or internet: it was a hardware play but with an internet interface. Niki has extraordinary analytical skills and had done modelling of companies like Atlassian and, more recently, web design company Canva, which also fit the pattern, to show what the trajectory of a successful digital startup should look like. He even applied it to the ultimate success of Blackbird:

> [Success is] not like twists and turns and panning for gold and reading the tea leaves. It either works or it doesn't. I'd be a bit more on the side of 'success is obvious from the beginning' versus the classic hero's journey of fits and starts and struggles and reaching the bottom of the Valley and things like that. Atlassian's a great example where success was obvious from the very beginning.
>
> People create the hero's journey. There's an empty bucket for 'adversity' and an empty bucket for 'things that were a threat to our company'. But at least from my point of view, Atlassian was a smooth, unstoppable journey upwards.
>
> If you ask the Canva founders, they would tell you that they encountered these hardships and these critical moments, but from watching Canva, I would say they did not encounter it. I've seen adversity in other portfolio companies and that wasn't what Atlassian and Canva experienced in my mind. They were unfettered successes from the beginning.
>
> As soon as Blackbird was started, it always just slapped us in the face that this was one of the biggest opportunities in the world. The opportunity slapped us in the face once we were started and I don't think from that point that Blackbird has encountered particular adversity [other than getting] our first fund up and running in the early couple of years.[97]

Niki was adamant that Zoox could do this too, so we met with its founder, Tim Kentley-Klay. Tim was a graphic designer in Melbourne and he'd rendered a fantastic picture of what the car would look like, however, he had never done a startup before and I was concerned as to whether he would know enough about how to build it, the market, or how to run a company. I was also concerned that he didn't seem to me to be interested in answering our questions and gave answers that seemed to indicate that he did not think they were real questions. For example, at one point I asked him how he was going to get an autonomous vehicle certified to drive on the roads. His response was that Google was going to blaze a trail and Zoox would just ride on Google's coat-tails.

While Niki loved him, I didn't want to work with someone who didn't seem open to help. Around the partnership table, nobody wanted to do this deal except Niki. Niki had the vision, he saw the opportunity and he saw something in this entrepreneur. I didn't want to invest in it, but I also didn't want to say 'no' when Niki clearly saw something that I didn't, so we did the investment. And wouldn't you know it? Tim went to Stanford, met their autonomous vehicle team, persuaded them to join the company, and built the thing. Two years later, Jeff Bezos and Amazon acquired it for about US$1.2 billion.[98] It was insane and shows that consensus is usually the poorest basis of good investment decisions.

I still believe in the importance of thesis-driven investment. In fact, it is a core tenet of Main Sequence (discussed in Chapter 6), but Blackbird has made many successful deals outside of the digital and internet field now. In a young venture system like Australia, the wider you can invest, the better it is for the system, so I acknowledge the greater good of the strategy. Blackbird is now the most successful venture fund in Australian venture history, increasing its worth from just under $30 million in 2013 to well over $500 million today, and with over $10 billion in net asset value across all its funds at its peak.[99] I think the reason it was so successful was that we stuck to our principles and vision, even if we expanded the remit for the types of startups in which we would invest. While the Zoox investment was a hardware play and outside of our market vision, it was unusual for another reason: very few deep-tech startups were being supported by Australian VC at all. Zoox not only broke the mould for Blackbird but also was an outlier in the system as a whole. This is where Australian VC needs to go next – investing in more of its own deep-tech startups – and there are positive early signs that this change is underway.

The next wave: shaping a unique Australian VC ecosystem

For many decades, Australia has lost its best innovations to investment overseas, from wi-fi to SunTech (discussed in Chapter 3) and many other inventions between. In recent years, we have made some progress through the second and third waves of VC: investors are no longer bankers with 70-page term sheets and chequebooks who don't understand the science; we seem to have seen the back of the showmen and ringmasters who promised to chaperone founders through the terrors of VC; and we've shown it's possible for Australian tech companies to list on both the NASDAQ and the ASX. Today's Australian venture community has an opportunity to shake off the legacies of past waves once and for all, to establish a strong scientific pillar of the Australian economy by supporting deep-tech startups. Blackbird co-founder Niki Scevak points to two deep-tech startups founded in Australia in recent years as examples of his optimism:

> Look at the newer crop of startups like [wi-fi chip manufacturer] Morse Micro, they've been able to raise capital and build an incredible team mostly in Australia. That would give me optimism. Sun Drive, which is solar panels using copper, is another one. All Australian and haven't had trouble building a team of the best photovoltaic people in the world in Sydney. I think there are plenty of examples to believe the case for optimism. I'm confident the newer cohort of companies will actually flip the equation.[100]

Powerfully, these examples represent the next generation of opportunities Australia has lost overseas in previous generations – wi-fi and solar panels – showing Australia does still have significant strengths here and the opportunity to build our own economic pillar from them with the right settings. In fact, Morse Micro has recruited to its board an original member of the wi-fi invention team, Dr John O'Sullivan, as well as Radiata alumni Professor Neil Weste and Dr David Goodall, and Sun Drive has Dr Zhengrong Shi on its board (whose story is discussed in Chapter 3), showing intergenerational goodwill for a stronger local innovation ecosystem. Both Morse Micro and Sun Drive are also portfolio companies of Main Sequence, discussed further in Chapter 6.

The digital revolution has made it not only possible, but also desirable, to start companies in Australia rather than in traditional innovation hotspots. For example, CEO of Freelancer.com and Escrow.com Matt Barrie said he deeply regrets moving his payments business Escrow.com into Silicon Valley a few years ago:

> The dynamics have changed a lot in the last decade. You can now have a global company with the management team here in Australia, as companies like Atlassian and Canva have shown. In fact right now, it's really hard to run a business in the US, and Silicon Valley is the most nuts place to try and run a business. When I bought the Escrow.com business, it was headquartered in Rancho Santa Margarita in South California. The

support staff were on US$31 000 a year. It was a good team but I thought it was in the middle of nowhere. You had to drive an hour and a half or two hours to get to LA. It's countryside suburbia. I thought we needed to be in the action, so I moved the office to downtown San Francisco with the tech industry who I thought would be our customers. Those support people now cost triple, they won't turn up to work, they want to go to a 'breakfast rave' before they turn up to work, the average tenure is 4 months, you can't get them to work because there's so much opportunity working everywhere and so many perks and free massages and meals and this, that, and the other – it's a stupid place to run a business. It's terrible. Now a lot of the selling is done over the internet anyway. All these startups in Australia you see that jump on a plane and go work from a motel in the Valley somewhere are just nuts. It's the last place you want to be trying to start a knowledge-based business.[101]

Reflecting on Australia's venture system when he launched LookSmart compared with today, Evan Thornley tempers his enthusiasm:

Australia is great now, versus 25 years ago when I started; there is an ecosystem where there was nothing. There was one venture firm, there were no accountants or lawyers who knew much, a few great people who helped us out a bit, but broadly there was nothing. Now there is absolutely a viable ecosystem here, perhaps a little one-dimensional in certain elements, but fabulous.

But let's not succumb to 'best in the southern hemisphere' mentality. A few years ago, I made the observation that one department of one faculty at the Technion in Haifa put more companies on the NASDAQ in one year than Australia had in its history. We had, at that stage, four NASDAQ-listed tech companies – I know, because I was one of them – and the computer science department of the Technion in Haifa put five on the NASDAQ in one year. There are two orders of magnitude difference between that ecosystem and ours.

We should celebrate the huge progress we've made, but if we are in the business of wanting to be globally competitive, let's understand we are pound-for-pound an order of magnitude behind Silicon Valley, and then disadvantaged for scale, but pound-for-pound two orders of magnitude behind Israel. Let's not pretend that the success we've had thus far is the limit of what we could achieve.[102]

Australia's innovation system is just beginning to form. The challenges we face now and the decisions we make will shape the kind of innovation system we build. Our ecosystem is still embryonic enough that we can define its DNA, both by learning from what – and what not – to do from other systems like the Valley, as well as identifying opportunities for Australia to mark out unique territory that will set us apart globally. Each wave of VC busting was the moment of epiphany for the next wave, in the same way it takes a near-death experience to catalyse success in a startup (discussed in Chapter 4). The first three waves of Australian VC have highlighted the following opportunities for us to shape this innovation ecosystem.

We need venture capitalists who understand deep-tech risk

When I was studying at Stanford in the late '80s, we'd meet venture capitalists engaging with the students on campus all the time. They did it partly because they knew these were the people likely to start the next big companies and they wanted to be well thought of, and partly because they needed and wanted to learn and get smarter about the science and technology that was disrupting their markets. In the 1970s, '80s and '90s, innovation was all about science discoveries like the semiconductor chip, the hard drive, the internet, telecommunications and wi-fi. You needed deep-tech. Put simply: digital startups are often ecommerce; software startups usually have unique code; and hardware startups invent new physical products.

In the pre-'tech wreck' 2000s, taking a company public and raising about US$50 million (enough for everyone in the startup to make a few million dollars) was the normal way of doing things. When we took Iridex public in 1996, we probably had more stock distributed among our employees than most Valley companies would, so we created a bunch of millionaires across our team, but not a concentrated group of investor billionaires who hadn't been part of creating the company from the start. Everyone who was core to the original team made enough money to buy a great house. These were simpler times with simpler expectations.

That's something Valley venture capitalists have backed off on today. They're still very focused on the market, but they're not spending as much time understanding the technology because innovation is no longer as closely tied to science and tech as it used to be. Now, the Valley has become more like Wall Street – focused on making investors rich – than about founders. That change happened in the middle of the telecom bubble. Just as the internet was becoming established, people on Wall Street realised that you could make enormous amounts of money from getting in early on these IPOs. During the tech bubble and the internet and telecom boom, Wall Street realised there was money to be made, so they jumped from US$50–100 million IPOs to US$1 billion and today we're seeing US$40–50 billion IPOs. You're certainly buying a lot more than one house with the proceeds. The trouble with that is it means that the people helping the entrepreneurs, the venture capitalists, are far more focused on the banking, the finance and the financial engineering of the exit, than they are the engineering of the products and the engineering of the company. What was wonderful about the Valley in those days is that when you worked with a venture capitalist, nine times out of 10 they were a successful entrepreneur, so they really could help you because they'd been through it themselves.

By embedding themselves in the research sector, old school Valley venture capitalists were getting an education in deep-tech-specific risk. When you see the time and money it takes to get ideas off the lab bench, into a prototype and then out to buyers, you get a sense for the TRL of an invention and that informs your investment decisions. In Australia, we don't manage risk in deep-tech by understanding science – we do it through layers of governance. Australian investors, including those in venture, can be a bit like Australian company boards in their reticence to take risks (discussed in Chapter 1).

There's a phenomenon that happens to most startups where, as you grow and become successful and get to profitability, you start to have something to lose; whereas when you start, you've got nothing to lose. You become more like a big company, which makes you afraid to bet the farm. A good deep-tech VC will understand that, frankly, until the company's gone public and is worth $1 billion, you should always be betting the farm, because if you don't, you won't compete with the big companies. That notion of 'calculated risk' is not very meaningful because there isn't much calculation needed. If you really want to be successful, you've got to bet big every time and hope that your science and technology are unique enough to forgive the multitude of sins that you commit when you make mistakes. If they are truly unique, and if your fundamental premise is that you've made something absolutely ground-breaking, you've got to believe that and bet big. Venture capitalists need to understand this and back you all the way.

While Australia's VC system has its origins with bankers, it has an opportunity to skip the Wall Street evolution of the Valley when valuations changed its DNA, because venture is still new here and isn't attractive to high finance yet. Instead, we can shape the people that we use to create it. We need to ensure Australian venture capitalists are connecting with research and redefining 'risk' in the context of deep-tech startups and investment.

Perhaps counterintuitively for a government agency, CSIRO took significant steps towards reimagining risk through the development of Main Sequence to manage the CSIRO Innovation Fund in 2016. By building the fund originally inside, and then ultimately alongside but independent from, CSIRO, the partners have access to scientific expertise and advice, as well as their own personal backgrounds in deep-tech. The fund will be slower to make its first exits than many other venture capital funds, recognising that deep-tech startups can take longer to reach their milestones. The fund is discussed further in Chapter 6, and I think is part of leading the next wave of VC in Australia.

Blackbird co-founder Niki Scevak echoes the folly of 'calculated risk', comparing it to 'due diligence', and says the best deep-tech venture capitalists are those who have walked the road themselves and have the same eagerness to learn as the Valley investors who spent time poking around the labs at Stanford:

> Startmate was about supporting 'technical founders with no business experience' versus 'businesspeople with no technical experience'. The same is true on the investor side – you need to have that empathy of seeing how a company is created and been through the journey of creating something. If you've created a company, or you've been through a few of those learning journeys, that is the better mindset for the investor.
>
> Due diligence, to me, are the silliest two words in seed investing. It's pretending to know by investigating and having the answers, versus just knowing that this is an interesting person with an interesting idea and let's go and explore this market and this product and this journey and see where we go.
>
> With Canva, the first investment was $250 000, and Blackbird has now invested $270 million into Canva. So, knowing that and looking back and saying, 'Would you invest 0.1 per cent of all the money you're going to invest in this company at this point in time?' Of course you would give it a shot in the hope that it does end up something like Canva where you end up investing that amount of money.
>
> You need an exploration mindset versus a lot of venture capital investors, who have a teacher mindset. 'I'm the expert, come to me for my advice.' An important question they ask when investing in a company is, 'How can I help the company?' Does that really matter? That's introducing your ego into the situation. At the opposite end of the spectrum is the student mindset: 'What am I going to learn from this person? What am I going to learn from this idea? From this company?' Always putting yourself in the student mindset will lead you to doing something new.
>
> You should, at some fundamental level, be investing in something that you know nothing about. Because if something is new, no one knows anything about it at certain points in time. There's obviously some people who know something and so on, but at a general level, you should invest in something you know nothing about – if everyone else knows nothing about it as well.[103]

Former CSIRO Chair Simon McKeon said there is no need to let not being a scientist stop investors getting behind early stage, deep-tech startups.

As a fairly prosperous and large-ish nation, we're still wet behind the ears when it comes to investing in early tech. If, as an investor, you're someone who doesn't know a lot about the particular technology field, the point is there are people who do, so get out of your zone of comfort a little bit and get to know these people. Yes, there's going to be an element of trust and relationship, and not everyone's going to have a PhD, but ultimately, having a preparedness to get serious about what's a really important part of the global investment market, which Australia doesn't seem to have its fair share of, is doable. A lot of people are just timid and say, 'Too hard, I'll let others do that.' There's no need to say that.[104]

He drew on his decades of experience at Macquarie Bank to suggest that having a strong risk function can actually liberate decision-making. He said former Macquarie CEO Allan Moss used the phrase 'loose tight'. The 'loose' part meant the best people were hired, and were trusted to go out and be successful. The 'tight' part meant they were really clear about their risk management limits, so no one took on bigger risks than Macquarie could handle. As a result, the risk function was worshipped inside Macquarie. The risk team continually questioned themselves so they weren't unnecessarily extensive and bureaucratic, like so many other corporates, but instead made sure the risks really were relevant and not as extensive as others might think they need to be.

When Australia's Future Fund moved from investing in venture to co-investing in startups as they led-up to their IPOs, CEO Raphael Arndt said the change gave the fund's risk and governance teams 'a series of heart attacks':

Unlike when you do an infrastructure asset or a buy-out investment, there's no due diligence pack and there's no accounting firm report on the financial statements and the accounts, and there's no legal support or report; maybe there is on the core IP, but that's about it. You're really backing the team of people and the skill of your fund manager.

We recognised that and decided to change our process to develop a cut-down, short-form checklist approach, so that our risk experts understood what was expected of them, which was different to what was expected of them in other sectors. We also actually built the process so that we could go from, 'Here's an idea, do you want to do it?' to 'Here are the dollars', within a week. In fact, in one case, we had to do it in three days – and we did. There's no way we could do that with proper governance on an ad hoc basis – we actually had to build a robust strategy and process and governance model that worked with appropriate delegations and so forth.

The cheque size was pretty small for us; we're writing $20–$50 million cheques in these co-investments, which, in a $200 billion fund is still meaningful, but isn't going to blow up the fund.[105]

Hostplus Chief Investment Officer Sam Sicilia said when Hostplus is looking for an investment, previous failure is a good sign:

> If somebody starts a company, and they have a big idea, they're founders. But if that company fails, not because of their misadventure or negligence, but for other reasons, then why wouldn't you pile in with your ears pinned back when the same founders come along again? They've already learned lessons playing with other people's money. You, as the investor that comes in for the second round, at least benefits from the fact that they're unlikely to make the same mistakes again. They've faced adversity and they've learned from that, and they're resilient. They picked themselves up and they're giving it another go. Of course, the first thing we do is reference checks in this country, to make sure they've had no previous failures. And if they have, we look at that negatively. I think it's going to take a massive effort to change that mindset, because it's generational. It has set in over 2-3 generations already. Changing that is not going to be easy.[106]

We must ensure the examples led by Sam, Simon, Niki and others in leaning into deep-tech risk are embedded in the DNA of our deep-tech venture system as it grows.

We need to tap the full potential of the ASX

When Intel was created in 1968, there was no NASDAQ. There was no market for a tech company to go public on. The New York Stock Exchange (NYSE) didn't want to know about tech startups – they were considered too risky. By 1971, the volume of wealth creation coming out of the Valley and its emerging tech sector caught the attention of the National Association of Securities Dealers (NASD), who set out to capture this wealth by launching the NASDAQ in 1971 to be the world's first electronic stock market. It was specifically designed to be different to the NYSE in recognition of the dramatically different businesses being created in the burgeoning tech sector.[107]

The more I got to understand the Australian venture scene when I returned in the early 2000s, the more I wondered – could the ASX go on the same journey? Do we need a tech version of our stock exchange? Australia tried something along these lines in 1984 to make it easier for more startups and small companies to float, although this was for startups in general, not just for deep-tech, as the NASDAQ is. It was called a 'second board' and between 1984 and 1987, 243 companies were listed, collectively raising $1.4 billion and with a market capitalisation of $3.8 billion. It outperformed the ASX main board in share performance in the 3 years leading to March 1987.[108] Unfortunately, just like

Australia's venture community in the 1980s, the 1987 stock market crash caused significant losses to the second board, from which it never really recovered and it was closed in 1996.

For years, Australian tech investors and companies saw an IPO on the ASX as an exit of last resort – they said it had no liquidity, no tech coverage, and nowhere to grow – so instead they always had their sights set on Silicon Valley and the NASDAQ. And for a while that was the right move, as it meant accessing the largest markets and reaching the highest possible capitalisation. However, since the tech wreck increased regulation in the US, in particular Sarbanes-Oxley (SOX) compliance introduced in 2002, the cost of going public has been driven up dramatically. When we took Iridex public on NASDAQ in 1996, it cost a few million dollars. Today it's more like US$10–15 million because of all the legal, compliance, risk and other costs involved. That's significant for companies that find it hard to carry the ongoing regulatory burden. Startups went from an IPO threshold of US$50–100 million in revenue to now needing US$200–400 million in revenue, which has shifted the market to the big end of town. The ASX doesn't have that huge compliance burden, so smaller companies can go public sooner, raise more capital in a public market and then grow into their market capitalisation; this is why I think the ASX is one of the great unfair advantages that Australia has.

After taking Arasor public on the ASX in 2006, I talked about it to a number of other interested Australians, like Teresa Engelhard at Jolimont Capital who was Chair of RedBubble and Matt Barrie who is CEO of Freelancer.com. Over the next 10 years, both subsequently listed on the ASX, with Freelancer.com completing an IPO in November 2013 with a valuation of $1.13 billion and Redbubble completing a $40 million IPO in May 2016. In May 2022, CSIRO spin-out Chrysos debuted on the ASX. In the lead-up, it raised the most money of any IPO that year so far at $183 million. This company defied gravity, pulling off an IPO despite a war and economic instability. Reflecting on the Freelancer.com IPO nearly a decade after it listed, Matt champions the ASX as 'not being a hassle', saying 'yes, there's a complexity about being a publicly listed company, but nowhere near the US complexities. It's fairly straightforward.'[109]

Similarly, Soprano Design founder and CEO Dr Richard Favero said of the ASX today that it would 'do what you need it to' and that 'the market here is as buoyant as it would be on the NASDAQ'. He nearly took Soprano public on the ASX in 2016, but on the morning of the float, the results of the US election started filtering in and the market stumbled on the news of the election of President

Donald Trump and all 23 companies lining up for the day had to hold off. He said Soprano was lucky that it didn't have to float; it was more about creating options for their investors, and the company hasn't thought about doing it again since.[110]

Unfortunately, since the GFC, the Australian investment community can still be nervous about tech and its vulnerability to shocks. There have been some notable exceptions, like Arasor, Redbubble, Freelancer.com and Chrysos, which I mentioned earlier, but I think the appetite still exists, if you can find the right market makers and the right analysts who want to learn the tech market. That could supercharge the opportunity of the ASX and similarly supercharge our own economy as companies grow here instead of relocating to appease analysts overseas. CEO of Looksmart, which backlisted on the ASX after its NASDAQ debut, Evan Thornley thinks the ASX's potential is already blossoming:

> It's fabulous how the ASX has developed now as almost the global mid-tier board, for the folks that aren't quite big enough for NASDAQ or the FTSE, but don't want to be on some second board in the UK or elsewhere. You've got, firstly, Australian companies that have a viable, credible, serious exchange on which to list and secondly, Israeli, European, and American companies listing on the ASX because it's a credible board, but it can cope with a smaller market cap than the NASDAQ or certainly than the FTSE. That is a terrific development.[111]

We need scientist CEOs

The first time I met with the Aussie founders who David Cannington had assembled in the Valley, keen to go courting US venture capitalists, I looked around and saw a motley crew. They were a rag tag bunch of (loveable) misfits to be sure, but as someone who had pitched to the prestigious and rigorous Sand Hill Road venture capitalists, I could see that there was no way they'd be taken seriously. And yet it's the kind of characteristic that I think is uniquely Australian. Whatever our version of 'innovation nation', Silicon Valley, or whatever we want to call it, it's going to be uniquely Australian in the same way those entrepreneurs were – and still are. Spreets founder and CEO Dean McEvoy agrees that Aussie founders still don't go down well in the Valley, but argues we should defend our way of doing things and instead re-educate the Valley:

> Every time you take an Aussie founder over to the US, you've kind of got to beat the Australianness out of them so they actually learn how to sell themselves. But I think it's actually a better way if we educated the world that where Americans over-promise and under-deliver, Australians under-promise and over-deliver. I think it should be just better known that when you hear something from an Aussie, it's more likely to happen

than not. So rather than trying to make us become these American-style pitch monkeys, we should do a better job of educating the world that, when an Aussie said they're going to do something, they're going to over-deliver.[112]

Replacing a scientist–founder with a professional CEO used to be the first step in the Australian venture playbook in the first wave of venture. You can't use a traditional investor lens to assess if an entrepreneur can run or create a science-driven innovation enterprise. If we can evolve our thinking about what the characteristics of a founding CEO or entrepreneur should be, particularly in science, it's an opportunity for Australia to be different to the Valley. Australians have a natural rebellious streak and often ignore the establishment, which is essential to doing things differently. This is also a weakness in corporate Australia where boards like to replace a CEO every 5 years, often from outside. Every company and CEO should be able to grow their own successor. US market data shows the best CEOs typically have a tenure of 10 to 15 years,[113] a surprising statistic. It took Atlassian 13 years to exit and it enjoyed the same CEOs through all that growth and change. Other than Alan Joyce, who was planning to retire from Qantas after 11 years until COVID-19 hit,[114] it's very rare for Australian CEOs to survive more than five. This won't work for startups and it's perhaps ironic that change and innovation can be best managed by a long-term CEO – if they act like a founder. Founders bleed for the company; professional CEOs bleed for their bonus.

While Blackbird and other third wave funds were all about the founder, the reality was they got better at identifying founders in digital startups and execution plays, but they didn't really work it out for science and deep-tech startups. This is the opportunity where I see most promise in the last few years, and I explore this in the next chapter.

Australia's future venture solutions

After three waves of steadily growing venture in Australia, I think our next wave has the potential to be the one we ride towards a true innovation ecosystem. The first wave swamped our founders under complex contracts and dumped them in favour of more seasoned CEOs. The second wave made progress in simplifying term sheets and changing the model of a venture capitalist from a banker reviewing a mortgage application to a coach and partner helping their founders grow companies. The third wave deepened the coaching and networking expertise of partners by drawing on Australia's network of founders to be mentors and investors to the next generation. If we can design our next wave of VC to be led by partners who understand – and embrace – deep-tech risk, who see opportunity in the ASX to grow jobs and industries here in Australia and who seek out and coach scientists to become our next generation of CEOs, then I believe we will have enough momentum in this wave to carry many, many more companies across the Valley of Death.

In order to get better at innovation we need more shots on goal. If one in 10 VC deals is the fund maker, then we need to get comfortable with the other nine not being successful. We need to celebrate those failures as learnings that will ultimately make us more successful in the future. If 'failure' in the traditional sense brings us closer to success in the entrepreneurial sense, shouldn't we count those, too? Are we measuring the right things on the ASX, or the NASDAQ, or do we need a new market metric? And when we fall short on these metrics, is it an indication of true risk, or are we measuring something entirely different?

Blackbird co-founder Niki Scevak believes we've made great progress in shaping our own innovation system in recent years, largely because we haven't had to deal with the firehose of innovation that is now drowning Silicon Valley venture capitalists' ability to be innovative or agile in response to the Silicon Valley ecosystem. Instead, Australian venture capitalists have backed the Australian startups that have been able to thrive in our inhospitable conditions thus far:

> At Blackbird, it was clear to see that the world comes down to these few generational companies [like Atlassian and Canva], so let's make Blackbird about investing hundreds of millions of dollars in those generational companies. It was easy to do that versus in Silicon Valley where the rules of the game are hardened and there's so many companies that are getting started every day. In Silicon Valley, it's easier to not question the industry norms versus somewhere like Australia where it's easier to question those norms.
> \   The Australian ecosystem has grown enough to have its identity and if you were to go back to that old banker way of doing it, the 'host' would reject the incoming fund, they'd be very quickly named and shamed, or cut out of the ecosystem with bad behaviour. I would say it'd be pretty hard for that to happen.[115]

Cyclical financial crises have flattened each wave, with successive waves of venture emerging stronger, but losing time and startups in each recovery period. Few Aussie funds have survived a full cycle of boom to bust to next boom. Aussie investors in VC firms haven't stayed with funds long enough to see light at the end of the tunnel, so recessions like the oil crisis in 1987, the dot com bust in

2000 and the GFC in 2007–08 generally killed each wave of Aussie VC. Our financial industry didn't have the maturity to really understand where venture plays at that stage and, even today, many funds don't realise the reason you have venture in a portfolio is that everything else is tied into market cycles, whereas venture is not. When the stock market goes up, housing is linked to that rise, interest rates go up and the dollar goes the other way. Everything is linked. But venture has its own time scale. You can have amazing exits in the middle of terrible situations, so the volatility tends to flatten out.

At the time of writing, a wave of economic tightening may determine whether Australia has developed the maturity to understand that venture operates independently of market and economic cycles. After two decades comparing Australian and US venture ecosystems, Tristen Langley suggests Australia needs to see through full investing and tech 'cycles' before we can expect VC to survive our risk-averse culture:

> Australia's in the formative stages of seeing what a downturn-upturn-downturn might look like, whereas US investor groups have seen that cycle over and again. Even the US angel investors have lived through – and beyond – pandemics to understand the ups and downs in the investing cycles and the tech cycle. If you look at the waves of semiconductors – shifts in materials, shifts in market leaders from Intel to Nvidia – you see innovations coming and going in waves. I think America has cultivated its resilience, whereas Australia hasn't seen that many cycles and we just don't have as many examples of companies emerging from the downturns.[116]

This is also reflected in our ASX, where our strongest companies have been in those positions for a long, long time, without any natural attrition from innovation displacing the big players. Even if we had seen companies displaced by innovative disruptors, our low appetite for risk would be more likely to characterise them as failures rather than as a natural market evolution.

However, there are some parts of the Australian investor landscape that have developed maturity from navigating this terrain for a while. Former CSIRO Chair and former long-time Macquarie Banker Simon McKeon said Macquarie has always taken the approach of 'looking for the next wave well ahead of it happening' and understanding 'the importance of seeing through the cycle, looking way ahead'. Simon said Macquarie found recruitment was also best done with a view to the longer term, rather than reacting to economic cycles: 'We've done our best recruiting when the Wall Street subsidiaries in Australia were saying, "Ah, we're in the middle of a slump, let's get rid of people." We picked up some brilliant people during those slumps, even if there wasn't a huge amount for them to do at the time.'[117] CEO of Australia's Future Fund Raphael Arndt concurs that rounds of layoffs tend to be good for innovation: 'Venture seemed to do quite well after recessions. The hypothesis for that was, people with good ideas lose their jobs and go to start the business they've always wanted to start, but never had the guts to.'[118]

If we can push past our instincts to hunker down when economic conditions tighten, then we could be on the cusp of nurturing the next generation of entrepreneurs – or, as we could call them, scientist CEOs, discussed in the next chapter. We have an abundance of reasons to believe Australia's Valley of Death is no longer the inhospitable zone or fatal leap it once was, but, as with any precarious crossing, the trick is to hold to our course – and not look down.

3

Championing scientist CEOs

When Silicon Valley was formed in the 1960s, it took its name from the invention of the silicon chip that was powering the tech-giants establishing themselves in the area. Those companies are why Silicon Valley has been better at recognising the model of the 'scientist CEO', like Gordon Moore, Andrew Grove and Robert Noyce of Intel, or Bill Hewlett and David Packard of Hewlett-Packard. But as investors in the US have become more financially oriented and less grounded in science and technology, they are, ironically, at risk of reverting to the early Australian VC model of planning to replace a science founder with a professional CEO from the moment they invest, which rarely works out well, as discussed in Chapter 1. This shift away from supporting deep-tech in the Valley creates an opportunity for Australia – but a perishable one, as nations begin to reinvest in science-driven innovation, as mentioned in Chapter 2.

Technology companies are led by visionaries, but today, not many tech companies are run by the visionary scientists who invented the tech that put the company on the map. Apple was, for the most part, led by Steve Jobs, but it was the genius of Steve Wozniak that drove so many of Apple's inventions. Elon Musk is the charismatic leader of Tesla, Solar City and SpaceX, but he is not the inventor of the electric car model that sets Tesla apart, the solar panels that drive Solar City's success, nor the rocket science behind SpaceX's innovations. Mark Zuckerberg is the face of Facebook, but he has been famously sued by others who lay claim to the original coding that turned Facebook into a zeitgeist. This is not a criticism of those leaders, but I believe we have an opportunity to reap numerous benefits if we support more original founders to be long-term CEOs, instead of relegating them to 'Chief Scientist' roles or running them out of companies altogether.

Chair of Australia's Tech Council, as well as of Tesla, Robyn Denholm said Australia needs more leaders who can take our wealth of good ideas and growing venture to succeed:

> In Australia's startup space today, there is no shortage of funding for good startups, there is no shortage of ambitious founders, but there is a

> shortage of good operators to take the ideas and the founders and create sustainable great companies. I do think in the startup space, Australia is doing well and punching above our weight, if you look at the absolute number of companies out of Australia and their market capitalisation, as well as the impact.[119]

Australia can embrace a new model of CEO – the scientist CEO – and develop a uniquely Australian, and globally successful, model to lead our businesses and inspire future generations. I grant you this notion is often met with disdain, but I remind people that the notion of female founders was also met the same way not all that long ago. We can learn from the shortcomings of the Valley to leapfrog their evolution and create a better model. This chapter charts the course forward by answering: why we need scientist CEOs to help address the lack of deep-tech companies in Australia; why scientist CEOs are more likely to grow a market rather than just steal market share from competitors; why we need scientist CEOs to solve today's increasingly complex challenges; and how embracing the scientist CEO model can create a more diverse and inclusive corporate leader in Australia, bringing the many benefits that flow from diversity. Australia has plenty of successful corporate CEOs, but we are not talking about creating a new economic pillar from the kinds of companies we already have – that's why we need a new model of CEO to lead them.

Create more deep-tech companies

The current crop of digital entrepreneurs in Australia is doing what I hoped I could inspire scientists to do by leading CSIRO – showing that scientists can also be great business leaders. They are ignoring conventional wisdom and not allowing venture capitalists to tell them how to run their companies or bring in seasoned businesspeople, but instead writing their own rules and delivering exceptional outcomes. In Silicon Valley, scientists found, lead, and IPO great companies – nothing stops them. They can create the economic pillar we have lacked for so long.

Economic complexity

Australian digital and software startups received a significant increase in investment following then-Prime Minister Malcolm Turnbull's 2015 National Innovation and Science Agenda (NISA) (discussed in Chapter 6). There is more venture capital flowing in the market now than there was before 2015 and a lot of VC firms operating now that didn't exist then. Chair of the Australian Tech Council Robyn Denholm said that our success in digital and software goes even further back: 'In the last 30 years, we have created 100 Australian tech companies worth $100m plus. More than twenty have gone on to become unicorns. They include great companies like Atlassian, Canva, Afterpay, Wisetech, Seek, REA, Airwallex, SafetyCulture, Gol, Cultureamp and most recently Employment Hero.'[120]

Despite that massive growth in digital and software, we aren't seeing as much interest in the rest of science investment – hardware startups. This leaves Australia significantly exposed, both through a lack of complexity in our economy and in our workforce capability. Being a hardware guy, I've always thought that when you really know what you're doing, you do it in hardware; when you don't know what you're doing, you do it in software; and when you really don't know what you're doing, you do a web app. It's a cheap joke and it misses the truth that software and hardware are inextricably linked in a dance together. Software will go through periods of leading innovation and growth and then a breakthrough will change the music and hardware will be in the lead again. Breakthroughs in software enable you to do amazing things with your hardware and vice versa. For example, Intel's multi-core chip enabled whole breakthroughs in data centres and drove internet dial-up levels that we couldn't have imagined before; Microsoft's Office suite democratised software. Managing Director of Microsoft Australia and New Zealand Steven Worrall said that while the pandemic has made online collaboration tools 'more understandable' and 'broken down some barriers', they are actually old tech:

> Artificial intelligence, machine learning, Internet of Things – these new technologies are readily available through cloud computing. What possibilities does that now create for companies to do some amazing things? As a nation we should be leveraging those capabilities, especially in industries where we already have a strong, global competitive advantage.'[121]

Because hardware and software are inextricably linked, getting trapped into investing in just one or the other is a mistake. You want to love and leverage them equally because they've both got to come to the party. If Australia is only investing in one side of the dance, we are going to be caught without a partner when the music stops. Hardware startups are still perceived by investors as higher risk and slower to deliver returns if they do succeed. This perception, which breeds hesitation, is not unusual. It's precisely the pattern of investment in Silicon Valley today, which has shifted from its early days of investment in hardware to now chasing the fast dollar in digital or software startups, which are all about speed to market: just execute, execute, execute. This is the polarisation that creates the first Valley of Death (see Figure 1, page x) between great science and great innovation. I think it's a false dichotomy to suggest deep-tech won't create jobs and companies. It may not create them as quickly or as cost-effectively, but I would argue the companies and jobs deep-tech does create are often higher-value and more financially sustainable – the simple fact is we need both, so let's be ambidextrous.

With the rapid maturation and increasing ubiquity of AI and the urgent need for rapid cybersecurity responses, it's easy to see why many people are 'dazzled by digital'. I've seen companies and governments alike invest extravagantly in digital technologies for answers to problems with their business model or their understanding of the market, but sometimes failing to see that digital is an enabling technology that can accelerate and amplify a solution, but not a solution in and of itself. The scary thing about digital is how easily people can be dazzled. If you've been an entrepreneur and built businesses, you know not to have unrealistic expectations about the transformative power of digital alone, because you would have failed if you did. The real power of digital is unleashed when combined with deep domain and market expertise, like genetic capabilities in agriculture and healthcare, or materials science in energy and resources.

To be fair to investors, hardware startups do take longer and have higher risk (discussed throughout this book), not least because hardware CEOs need to be able to invent their product, so they will burn through more capital in that invention process than a software company before they are market-ready. Semiconductor company Quantenna went along for years making no money, spending $100 million to get to a working chip, but when they hit it right and cracked the tech, suddenly the revenue went to the Moon. That's why people still invest in semiconductors, even though it's so much harder than an internet startup, as seen through the US government's CHIPS (Creating Helpful Incentives to Produce Semiconductors) Act and the creation of the American Frontier Fund.

Australia's world-class scientists can help us better understand and therefore lessen the risk of these investments, especially if they are the CEOs leading these companies. In hardware startups, you need a combination of market expertise and company commitment as well as science or engineering credentials. You also need someone with vision who can inspire a team to push through the tough barriers of invention. You've got to invent it, engineer it, monetise it, find a customer, and then find enough of them to grow a company. You go from idea, to feature, to product, to company, to market – and any one of those points is a failure point. It's a very different scenario to a digital startup like Instagram, where the founders went from idea for an app to billion-dollar acquisition by Facebook in less than 2 years.[122] It's an extraordinary story, but that's a very different team to what you need in a science startup.

See more company opportunities

The returns may take longer, but if you can actually invent your unique product, you'll have an asset that acquirers will value even if the company doesn't work out, and your investors will still get a return. At Southern Cross, Bob Christiansen and I invested in Mesaplexx, an Aussie hardware company trying to commercialise superconductors into the telecommunications network to open up more space for the growing number of mobile phone users.[123] It was extraordinary technology from the University of Queensland to invent a high temperature superconductor. Superconductors generally only work at very low temperatures like absolute zero (about –270°C), but the university had figured out how to make one work at a warmer temperature (say, –150°C), which doesn't sound very warm, but it can be more easily achieved than absolute zero by using a cryogenic pump. Despite the genius of the invention, we still faced the scientist's conundrum: what do we do with it? We worked with the Mesaplexx team and VC fund Uniseed to figure out what could be the 'killer application'. Mobile technology was exploding at the time and we were running out of spectrum and bandwidth. The superconductor was an amazing filter, enabling channels to be placed much closer together to solve the bandwidth problem. There was even a place in the network where the traditional filter was installed so we could target a drop-in replacement, which is an ideal way to innovate because you have a really clear target with high market certainty, a perfect fit for our product requirements document, or PRD, a concept explained in Chapter 5. The only detail we needed to resolve was how on Earth we were going to put a cryogenic cooler up a telecom tower! It had never been done before. Believe it or not, the company did it in partnership with a Finnish telecommunications company and actually ran live traffic through it, something a carrier would normally never risk, and it performed remarkably. The problem was the perception of risk with such a device and that there may be other ways to do it. The arrival of 5G did eventually replace this need, but it happened much later than anyone had expected. The company didn't work on its own in the end, but Bob amazed all of us by figuring out how to sell the company, the team, the product and the IP for a good return to Nokia-Siemens in 2014 who valued it as an enabling platform technology for their network.[124] This would never have been possible without a scientist CEO who deeply understood the tech and the risk but also the platform nature of the invention.

Wave was an even more extreme example of a visionary scientist who had no idea how to make their idea real. Southern Cross helped create and then invested in Wave during 2010. Wave was an idea to develop semiconductor chips with asynchronous logic; that is, chips without a clock. Processor chips in your laptop have an internal clock that sets the pace at which each instruction is executed before the next. When you buy a computer, you always care about the clock speed by looking at the processor – Intel's current 11th generation processor is 5 gigahertz. It would mean the speed of chips would no longer be limited by the clock. It's a little bit like a quantum chip in that rather than processing instructions linearly, one after the other, it processed them in parallel with a massive increase in speed, representing a major paradigm shift in design. It was very out of left field and basically uninvestable under normal VC rules because it was an innovation in semiconductors when everyone was obsessed with the internet instead – never mind that semiconductors actually power the internet! Because we had invested in Quantenna, we knew that a great science innovation with the right market vision can break the paradigm. Even our investment partners at Tallwood VC, who are semiconductor experts, were leery of the concept. But Dado Banatao, founder of Tallwood and a number of great semiconductor companies like Marvel, could see the market vision and potential for shift. Quantum chips weren't real then, but this could jump ahead of them, using existing silicon practices.

The development of the chip brought in scientists from CSIRO and what was then NICTA, including their Chief Technology Officer Dr Chris Nicol, who sort of became the founder of the company because he was able to figure out how to make it real. Pete Foley was an engineer who worked at VC firm Tallwood and became the CEO – with no business experience but a deep understanding of the science and the market. I mentored Pete for years on this project, both as his investor and board member. The team made this chip run so fast that it melted the silicon; it literally melted glass. It was extraordinarily successful because it created an opportunity to use a conventional silicon chip architecture to do something that would get close to quantum. This company was spoiled for choice in applications: it could be a processor for PCs or handsets or virtually anything. It was a new platform with the classic platform challenge of where to focus. You have to pick just one thing to start with and focus on delivering it before you can leverage the platform, as discussed in the next chapter. Wave picked several opportunities before it ultimately hit the right one. The killer app turned out to be 3D graphics processing for virtual reality.

The company was a unicorn, financed with a $1 billion independent valuation. It ultimately was acquired by a Chinese company through a range of subsequent business deals, but the technology itself was never the problem. This story became one of many that catalysed the US government's creation of the CHIPS and Science Act in August 2022, as well as the creation of the not-for-profit deep-tech American Frontier Fund in June 2022. The shame for Australia was that before the company was off the ground, there was a leadership change at NICTA and they exited the project. Its new CEO, Hugh Durrant-Whyte, had a background in digital, not hardware, and didn't continue with the project. That's completely understandable, especially as a new leader with a new program to implement. But it does go to show how specialised your knowledge needs to be to see the value of an investment in a science field that isn't your own. Also, that decision forced Chris and others to make a tough decision to leap into the void instead of play it safe at NICTA, which changed them forever into entrepreneurs.

Disruption protection

A software company has a harder time protecting their IP than a hardware company and competitors will be on their doorstep that much faster, because you can't patent code. If your only advantage is being fast to market and good at execution, that's not an uncommon skill, so you are more likely to be disrupted. As soon as you go into market with an internet deal, competitors pop up all around you far more quickly because the barrier to entry is low. For example, Blackbird invested in Shoes of Prey because it was digital-meets-hardware. Shoes of Prey used a fairly simple internet platform to sell shoes online, but it had a quite unique way of manufacturing the shoes, customising them to each individual customer, so the business model was a mass-customisation play. The company had some very clever technology in the backroom that got us interested, but within weeks of it launching, three competitors popped up. Shoes of Prey was successful for a decade before the market caught up and it was priced out by competitors who copied its business model.[125]

This is a big part of Australia's innovation dilemma because investors here 'know' quick wins come from backing execution plays, like digital and software startups, led by experienced CEOs. But to grow our economy, to differentiate ourselves in a competitive world and to build whole new industries and markets, we must create more scientist CEOs who can understand and de-risk companies so investors will back them to win. Software and digital plays are currently leading the dance in Australia, but we will be caught out when the music changes if we don't address this imbalance. We need experts in hardware leading companies just as much as software.

Grow the market with missionary CEOs

The kind of innovation that governments need in order to grow their employment base is additive to industries and markets; it creates new value and growth. 'Innovation' is sometimes referred to interchangeably with 'disruption', but while it can be disruptive to eat someone else's lunch, it's not very innovative. More simply put, there is a fixed amount of cash being spent in any given industry. If you want a portion of that cash, you can take market share from competitors by becoming another provider; you can insert yourself into a transaction to skim off cash that is paid to others by becoming an intermediary; or, ideally, you can grow the market by adding functionality or by creating a superpower that no one else expected. Growing the market takes vision – it means seeing a future that no one else sees, which is precisely what science does. Science not only forecasts the future – it also makes it happen.

In Australia, I think part of the reason we don't have many deep-tech CEOs is because our scientists are deeply uncomfortable with charging for their expertise, which goes back to the 'cultural cringe' innovation barrier discussed in Chapter 1. Australia needs more purpose-driven 'missionary' deep-science startups because, historically, Australian entrepreneurs have been 'mercenary' CEOs who struggled with customers paying a price for a product that is only a fraction of the value it creates for the customer. It feels like leaving money on the table. Across many parts of this country, we have a 'finite-pie', 'zero-sum game' mentality. We're more like traders who compete for money than entrepreneurs who create value. The opposite mindset drove Silicon Valley in its heyday: don't worry too much about how big your slice of the pie is, because if you help make the pie enormous it won't matter. The venture community was keen to back big ideas because of the potential for them to be huge. When the idea is in the realm of 'it could be bigger than I can possibly imagine', it doesn't matter how high the risk is. Of course, you're probably going to fail. But if you are successful, it makes up for everything else. In Australia, we don't think that way. Regardless of whether you are working on quantum computing or something less complex, your product still has to create enormous value for the customer in order for them to buy it and you're going to have to leave that value on the table for them. If you try and take all of it for yourself, you won't change anything. That's the price of change.

GraphAir

In 2015, CSIRO set up the ON accelerator to support scientists in commercialising their brilliant ideas from CSIRO and Australian universities (discussed more in Chapter 6). Very few of the scientists who participated in the program had actually engaged with any customers in the real world. One of my favourite examples that came out of this program was a CSIRO team who had invented a new way to make graphene, a single layer of carbon atoms arranged in a hexagonal lattice structure, and called their technology GraphAir. Graphene is exceedingly expensive to make, but incredibly versatile, and the team found a significantly cheaper way to make it using soybeans.

They initially thought they'd sell their graphene by being the cheapest on the market, and they worked on this model for many weeks before we explained that competing on price is a race to the bottom and almost always shrinks the market. Although price elasticity can grow a market in some cases, who wants to be the lowest cost provider of anything? Innovation thrives when the customer no longer cares about the price of the product, because it is so compelling that they want it and they understand that its value isn't anything to do with what it costs to manufacture. If you solve high-value problems, you'll never argue about price with customers. Instead, you'll always be arguing about delivery. When can I get it? How do I get more?

If you create value for your customers – tech companies do this well, and deep science companies do it amazingly well – the customer transitions from having an adversarial relationship based on buying a product for the lowest price to being part of the fan club (discussed in Chapter 5). That's priceless because it will give your company the exponential kickstart it needs to get going. Years later, when you're a real company with lots of customers and products, while still true it will be less important, because your brand and your reputation for quality will start to offset the necessity to give more value to the customer than to yourself. It's like a threshold problem: you have to get over the threshold to become real but that costs you quite a lot in value initially.

So, we rejected the graphene team's first business proposal. They were brilliant scientists and it really knocked them for six to fail at something. We talked to them about how everybody fails this test and, in fact, that failing it is the only way you succeed.

They tried lots of different markets and customers to find their unique value proposition. Ultimately, it was their altruistic sense of purpose that drove their decision. The one that really stuck was using this unique and cost-effective product to improve water filtration, because it meant they could take their solution to the parts of the world where it was needed most, but least able to be afforded. It was interesting and inspiring to me that they fell in love with the purpose of that product, in an effort to solve the problem that 70 per cent of the world's disease comes from unclean water. It's the number one killer in the developing world and they were passionate about changing that problem, even if it's not an application that is going to make them and their eventual commercialisation partners incredibly wealthy.

The other piece of the puzzle to finding the right application for an invention is not necessarily picking the absolute optimum application, but rather finding an application that will give you enough passion and purpose in your team to want to push through seemingly impossible barriers to make it happen. For me, watching that team have the epiphany that took them beyond the brilliance of their invention to instead start to worry more about who could benefit from it and how they could help the world was a magical moment. As soon as I saw that, I was convinced that team was going to be extraordinary.

When we announced the breakthrough invention and intended application, 150 companies from around the world jumped on them, infected with the same passion. Why was that so important? Because it's not as simple as making a filter, holding it over a bucket and pouring water through it. For it to really work, it needs a whole machine around it with lots of other mechanisms. It drove them crazy that it couldn't be simple! But they were willing to learn all of that really boring stuff from their prospective customers about how they could actually make a product, because they had passion for their purpose. I think that's a secret weapon of deep-science startups.

Individual motivation

Scientists know that if you want to invent the future, you'd better have something stronger than money motivating you to get up every morning and keep pushing through failure until you find your breakthrough. That's why scientists so often become missionary CEOs, driven to make the world a better place, as opposed to mercenary CEOs, who are driven by profit. CEOs who start businesses because they want to create something new and better don't always succeed; but those who are in it just for the buck almost never do. The fire inside your belly can sustain you through the ordeal, but greed alone will not. Steve Wozniak recalls the creation of the amateur 'Homebrew Computer Club' in March 1975 by many of the future leaders of Silicon Valley, writing that everyone 'envisioned computers as a benefit to humanity – a tool that would lead to social justice':

> Only big companies could afford computers at the time. That meant they could afford to do things smaller companies and regular people couldn't do. And we were out to change all that. In this, we were revolutionaries. Big companies like IBM and Digital Equipment ... didn't imagine how [hobby computers] could evolve.[126]

It's the vision of the future that no one else sees that drives so many scientists to try to start companies from inventions. Things not being right in the world really annoys scientists. When we see a product that's not right – that could be better, that could have greater capability or serve more people – we want to re-engineer it or redesign it to make it better. I left my first job at a company called Fibertek because it started moving into defence applications and I saw a different future for the products Fibertek was making. Founder and CEO of Spreets and Booking Angel Dean McEvoy said his passion for giving something to the hospitality industry was driven by growing up in a hospitality family and that, not only do scientists make missionary CEOs, they are the only kinds of CEOs he wants to work with:

> My first startup, Booking Angel, was an online reservation system for restaurants launched in 2004, using advanced speech recognition technology to create a phone/website interface. I got excited about this because I grew up with my family owning bars and restaurants, so it was actually solving a problem that my family couldn't solve, getting bums on seats. For me, that was the rush of building the business, the impact you have, and the people you influence.
>
> When we launched Spreets, serving the same audience, that passion really helped us, because a lot of people came into the group buying market when it got competitive and their reason for starting it was this successful business model and went and copied it. I understood that problem in a way that other people didn't and that allowed us to sell it better, it allowed us to scale the team better, build it better. We had competitors that were spending $30 million on media. We raised $800 grand, so we just did it all on the smell of an oily rag, but we got there.
>
> It felt amazing, but not amazing for the reasons that a lot of people think. They think, 'Oh, you made $40 million in a year.' It's like, 'No, the investors made most of it, they made 18 times their money in 6 months.' It's more, for me, about the impact you have, particularly on small business, I'm passionate about that, and the impact you have in building a company culture where people really loved working there. I don't think any entrepreneur that's really successful is motivated by money that much. Money is like the scorecard, that shows you had impact. So it was finally good to see some ROI on our efforts in terms of people actually finding what we did as being useful.
>
> It seems to be something that resonates with me, and it's also what I've seen in other entrepreneurs. When you can find the connection with why they're doing what they're doing, when it connects with them as a human or their childhood or something, they're so much more motivated to stick at it and they'll come up with much more interesting insights around the problem that mean they can attack it in a different way.

From an investor's perspective now, it's what I look for in companies and people. The 'sciencey' startups I've invested in tick the box of, is there someone in there who really cares about the problem in an authentic way? Like their 'why' behind it is not, I read a post on TechCrunch and it seems like a good sector. Or, there's a Gartner report that said these areas are in the magic quadrant. No. Never. Rubbish. It still needs the individual who will unfairly give a shit about this enough to never give up, to keep going at it when others will give up. You need to see that sparkle in their eyes, and that attachment to the problem that is irrational – because scientists are rational people, like X + Y proves Z – so any good scientists would look at the statistical likelihood of being successful in building a startup and go, this is a bad idea.[127]

Former CSIRO Chair and former Telstra CEO David Thodey sees the altruistic objectives of scientists and good business leaders as closely aligned:

The objectives of business and scientists are closely aligned because, to me, science and research have always been about gaining a better understanding of the world around us and business is about using that understanding to provide better products and services. It's about understanding the needs of markets or individuals, and then it is about using knowledge and innovation to address those needs. I think scientists are committed to better understanding the world around us and then to apply that knowledge to make the world a better, easier place to live. Obviously, there is a difference in that business has a commercial objective, but I would put to you that any good business or business executive always has a longer-term perspective. The short-term drive for profitability is a misdirected objective. The objective of a sustainable and successful business is to build companies that have long-term impact – there is a commercial imperative for companies to have a purpose and a sustainable impact on the societies in which they work, which is just good business. When I invest in a company, I look at the idea, at the business model, but I also look to the quality of the leadership, because that actually has a significant impact on my decision to invest – do they have values and purpose that will enable them to achieve their ambitions?[128]

Fortescue Metals Group Chair Dr Andrew Forrest shows a missionary-like approach to pushing through setbacks and failures:

I believe that if something is true and just and the world needs it, then the expense of your failure is nothing on the dividend of the success. It's much more costly to you, because you have personally failed and no one gives a toss. But if you keep looking at the bigger picture, that actually, if I were to succeed through attempting again and risking failure again, then the dividends to the world is much, much bigger than my actual loss and failure, then I'm going to go again. It's thinking in that wider frame that it's not about you, it's about what that contribution could be. That's why I stay determined.[129]

Chauncey Shey is the visionary founder of SoftBank China, a great entrepreneur before he became a venture capitalist (he was co-founder of Chinese global telecommunications infrastructure provider UTStarcom), and an extraordinary judge of character. Chauncey tries to see the heart of a person – what he saw in the heart of a young English teacher with no business qualifications made him invest the very first major money into Jack Ma's ecommerce company Alibaba and subsidiary, TaoBao. That first massive success bought him the ability to invest in much riskier projects, including when I first met him and pitched him on co-creating an Australian focused Renewable Energy Venture Capital Fund (REVCF), which we ultimately did with the backing of the Australian government in 2010, discussed further in Chapter 5. Through the REVCF, we backed a young visionary scientist to create BenAn Energy to enable safe high-density energy storage, and much later, my old friend and fellow physicist Dr Zhengrong Shi to create Sunman Energy to reinvent and democratise solar. Chauncey has a passion for scientists and inventors. If he can see the fire in their heart, he will back them, mentor them, and help them change the world – because he knows scientists have learned to learn.[130]

Motivate the team

There are usually two reasons that employees join you early on in a startup. The first is they have fallen in love with the idea – and a big, purposeful idea is important if you're trying to recruit really good scientists and engineers. The second is they have fallen in love with you as a leader. The second reason is because they can see that you absolutely believe that this thing is going to work in your heart and soul, it's the right thing to do, and it is going to change the world. That energy becomes infectious, bonding you together as a team and people follow you. If your team isn't absolutely committed, it won't hold together. Without the 'followship', the first time the prototype fails, the funding runs short, the customers change their minds, or any number of other crises, the whole thing will fall to bits.

Iriderm

At Iridex, we were trying to cure eye disease using lasers and so everyone in it was a bit of a missionary wanting to change the world. We had big established competitors, but when you are on a quest, you don't let the dragons deter you; rather, you focus on the journey. The doctors loved us because we were disrupting a complacent market that wasn't serving their needs, but it was clear that despite our naivety in thinking we could compete with 'Big Pharma', we really understood our market and customers. And those customers also wanted to change the world a little, so they supported us to intentionally spite their existing suppliers. We didn't have the budgets to wine and dine them like Big Pharma would have, but we sure made their work better, faster and more reliable.

The green laser platform was so unique we wanted to expand our market into other medical treatments, so we tried dermatology and created a new company called Iriderm to treat skin disease. It worked out well, but then we tried applying it to aesthetics. I did feel a bit of unease because aesthetics isn't like therapeutic medicine. The plastic surgeon can be considered more of a businessperson, with a focus on elective surgery. It was hard to manage that clash of DNA inside a small startup, where being united behind a single purpose is crucial. Even when we went public, the fissure remained. It's a great lesson in what not to do inside a startup, which is so critically dependent on pure DNA. We didn't quite get to the toxically divided culture of the Mac team versus the Apple team, but you get the idea. (I discuss the Iridex cultural clash further in Chapter 5.)

My missionary purism was tested when my young daughter came to my bedside in the middle of the night saying, 'Daddy, something's wrong.' I woke up and groggily walked her to the bathroom and turned on the light to find her covered in blood. In short, there's a fine artery connected to the ophthalmic artery (which I only knew because we were an ophthalmic company) that feeds blood to your nose and face. It's fairly common that this small vessel may push up to the surface and rupture, but it's very difficult to stop the bleeding, especially as an 8-year-old seeing panic on her daddy's face at 3 am. There is so much pressure in this artery that the bleeding will continue for hours. For a while, it ruptured every second day with disastrous results.

Two weeks later, after another dramatic bleed, we rushed into a famous aesthetic surgeon's lab in San Francisco and, using a certain company's aesthetic laser, he cauterised the offending vessel and healed the wound. As the surgeon finished the treatment, my daughter said, 'I'm lucky my daddy is a doctor,' to which he replied, 'No, you're lucky he is an entrepreneur and I am a doctor.' Our ophthalmic company had received thousands of letters from grateful parents whose children had been cured by our ophthalmic laser, but never once had I understood that the aesthetics laser could really treat illness. It was a wonderful lesson in humility and not being so opinionated about what matters and what can change the world. However, my earlier point stands about the rupture this caused within our company. Everyone inside your own startup has to believe they're on the same mission or it won't work.

Conversely, once you've been through all that and you start to get to the growth phase of the company, those events are what will make the company so successful. You've learned and you know how bad it can be so you'll never get arrogant or complacent about your success because you know it is a rare gift. It keeps you a little bit paranoid, so you keep managing, you keep reinventing, you keep driving. If you're not like that as the founder, people will see through you. When you're working in a startup, you're living together; you're a family. You can't put it on – you've either got it or you don't.

That's a question every scientist needs to ask themselves when they decide they want to try to go down the entrepreneurial path. If you're doing it because you really want to change the world and you really believe in your idea, then that passion will attract others. But if you're doing it because you think you're the smartest person in the world and you want to be famous on the front cover of magazines, your team will see through that, they won't gel and you'll end up with a team that's like you, all wanting to be on the cover of magazines – and that doesn't get anybody anything.

Quantenna

When Southern Cross invested in Quantenna, it limped along for about 4 years, not really generating any revenue. It gained a bit of customer interest from Swisscom, but not really anything big. The other venture capitalists were getting antsy, so they decided to fire the founder CEO and bring in an experienced CEO, more in line with the more recent Valley playbook. They brought in Dave French, the CEO of a big public company and a very successful CEO, but his culture was not Quantenna's culture. As I mentioned when discussing Iriderm's work with the aesthetics industry, you can't really transform the culture of a startup. You've got to start again with different people, but if you do that, you mutate the DNA. Eventually Dave and I approached the other investors at Sequoia to talk about who should be the CEO, which turned into a months-long fight.

While all of this was going on, I noticed one of Quantenna's key engineers, Dr Sam Heidari, was very quietly - with no authority but knowing it needed to be done - pulling everybody together and keeping the company on track. He didn't tell people what to do, he just had their trust and knowledge that what he said made sense, so people naturally followed him and the company kept going despite all of this chaos. He wasn't a founder of Quantenna, but that was when I realised he should be its CEO. Sam didn't want to do it, but he eventually agreed to do it if I would help him. I mentored Sam for years and, as I look back on it, I think I learned more from him than he learned from me.

Once Sequoia VC really got behind Sam as CEO, there was no stopping him. Sam took the company public, although he wanted to step out and bring in someone who had experience running a public company, like Dave French, to run it. We all said 'No! We're going to teach you how to do it. You're going to be great.' And he was - he went from engineer to CEO to public company CEO. It was an amazing journey for him, but I don't think anyone would have picked him if they didn't understand the need to think differently about what a scientist CEO can look like.

While 'followship' is essential, even the best team players are working for a pay cheque. When we took Iridex public in 1996, at least 20 per cent of the company was owned by our employees, if not more. Everyone who was core to the original team made enough money to buy a house. If you can imagine the joy on the faces of your loved ones when you pay off their mortgage or set up an education fund for their children, then that thought will carry you through all the tough times that inevitably plague every startup and keep you going right through the IPO. Again we see this play out in Steve Wozniak's experience, where he tells how he came up with 'the Woz Plan' to sell stock in Apple cheaply to people he thought deserved it and who wouldn't otherwise be able to have a share in Apple's success, because he didn't think it was fair that regular employees didn't get the stock options the executives got. Almost everybody who participated in the Woz Plan ended up being able to buy a house and become relatively comfortable.[131]

Australian startups don't share enough stock with their people and I believe increasing the prevalence of employee share option plans (ESOPs) is key. Historically, Australian tax laws haven't made this easy, although a number of improvements were introduced in 2015, around the same time as the National Innovation and Science Agenda (discussed further in Chapter 6).[132] We need to see them become more common now, especially as startups are in their early investment-raising stages. As Atlassian grew, Mike Cannon-Brookes and Scott Farquhar used the opportunity to give their people access to ESOPs so they could share with their employees, despite the expense of government regulations and tax barriers at the time that made ESOPs unattractive in Australia.[133] Both Main Sequence and Blackbird advocate greater ESOPs than are currently common, pushing for at least 20 per cent of the company's equity to be in the hands of its people in the first funding round.

This idea of 'paying it forward' is part of the DNA of missionary CEOs and of the Valley in its heyday. One of the Google founders, Sergey Brin, lived down the end of my street when I lived in Los Altos. He had grown up in that area and he loved the feel of the place. Sergey was worried that all the money that Google was going to create could gentrify his neighbourhood, so he bought a half a dozen shops and he turned them into memories from his youth. He created a candy store with the big jars of candy priced at 2 and 3 cents and sold in little white bags. He opened a toy store where the staff actually helped the kids put the toys together, and they'd all sit there for hours doing this and having coffees. It didn't make any money, but he didn't care. It was really important to him to have that 'old school' feel. For a guy that made so much money out of Google, he was much more concerned about keeping the neighbourhood special, having that uniqueness.[134]

When Google grew really big and took over Mountain View, which is where I lived before I moved to Los Altos, the Googlers were everywhere, wearing their Google T-shirts and riding the Google bikes. But Google told its employees: you're part of this society, you're part of this community, and you've got to give back. They did clean-up days and other community service days. You don't necessarily see that behaviour from Facebook or Instagram in the Valley today, but you do see it in some of Australia's tech entrepreneurs. Mike Cannon-Brookes' and Scott Farquhar's Atlassian donates to the Room to Read charity for children's education, among other philanthropic activities, based on its 'one per cent model', which means one per cent of its profits, equity, products and time are donated to charity, with employees given 5 days of paid leave each year to volunteer for non-profits.[135] Similarly, co-founders of internet design company Canva, Melanie Perkins and Cliff Obrecht, have committed $1 billion to create a charitable foundation.[136] It's something we need to nurture and encourage in our ecosystem as it grows – but if our founders are true missionaries, they won't need much encouragement, whether it's growing their market or growing their communities. It's wired into their DNA.

See multiple futures in an ambiguous world

Today's challenges are more complex and interconnected than ever before. CSIRO research has found that some of the United Nations' 17 Sustainable Development Goals are, in fact, mutually exclusive and that we will need to make uncomfortable trade-offs if we are to make overall progress.[137] New fields of research require cross-disciplinary teams to make progress, like quantum technologies and synthetic biology, while others are developing so quickly and with such broad applications that new disciplines like 'responsible innovation' are being created to cope with the pace and scale of change. The European Union, Australia and other countries are creating multi-stakeholder 'missions' in order to take on bold, ambitious goals (discussed in Chapter 8), recognising that diverse coalitions of research, government, industry and community groups will be needed to effect change in today's complex environment.

What kind of a CEO is best-placed to deal with unprecedented ambiguity, opportunity and choice? If you believe in personality types, there are some generalities that bear out in life. The Myers–Briggs framework suggests that the entrepreneur (ENTP type)[138] and CEO (ENTJ type)[139] personality types represent about 3 per cent and 2 per cent of US population respectively, with both described as rare. They're very similar personalities but for one trait: it suggests CEOs *judge* the world and prefer firm structures, where entrepreneurs *perceive* the world and are more open, flexible and adaptable. CEOs want clear KPIs and metrics; entrepreneurs are open to multiple possible futures.

I'd been an entrepreneur and a CEO many times by the time I became CSIRO's Chief Executive at the end of 2014. When I started, I really wanted CSIRO to have a global top 100 CEO as its Chair, so I approached former Telstra CEO David Thodey, who took on the Chair role at the end of 2015. In every sense, David is a very, very good CEO and that's not just my opinion: he ranked 61st in the annual *Harvard Business Review* list of global top 100 CEOs.[140] He exemplifies the Myers–Briggs ENTJ type, with a strong focus on KPIs, metrics, frameworks and management tools. David's special strength as Chair of CSIRO, an organisation of thousands of scientists who love complexity, was his ability to simplify. It's actually genius. Scientists can live with complexity, their brains can make sense of it, but a good CEO will look for the simple metrics and most important areas of focus to provide guidance and leadership. David said it was a challenge scientists face, noting:

> I think great leaders in the world – and I think scientists should be great leaders – have this ability to work through complexity to get to the truth, which is often very simple. Our world is very complex and many people become lost in that complexity. Great leaders help us understand what is important and break complex issues down to simple principles. Scientists and others can sometimes be too focused on explaining the complexity or the process of analysis they have used, rather than providing insight and clarity, and explaining the impact of their work.[141]

This 'process first, conclusion last' approach is something we tried to invert through the ON program, discussed more in Chapter 6. Entrepreneurs are a bit like scientists in this way; this is the dimension of their personality that's different from a CEO. Entrepreneurs see infinite possibilities and they figure out how to navigate all the angles. CEOs are often very good at making decisions, but then not very good at un-making decisions when they get them wrong. Whereas entrepreneurs can see all the possibilities and can make quick decisions, but they never lock them in. In their mind, they're always going to be testing that decision, asking themselves, 'Was I really right? Does the data support that? Or do I need to pivot?' That's the magic of the entrepreneur.

In a big company, you will drive your board and your management team crazy if you behave like an entrepreneur when you are the CEO. I know, because I did! Although, David generously told me: 'I admire and respect your rigorous and disciplined analytic mind – your background in science – you've had a very different career to me, and I really value that as you bring great insight.'[142] David and I brought completely different perspectives to CSIRO, and both the organisation and I have grown as a result. I always struggled being a CEO because I can't tell you how many times in a day I want to jump in and do it myself and change things. As a CEO you need to do the heavy leading, not the heavy lifting, but as an entrepreneur you must do both. I've learned over the years as I've become older to rein that in.

Myers–Briggs and other personality classifications suggest that as you get older, different aspects of your personality tend to evolve and you can start to develop them – which should give any scientist who doesn't fit into the Myers–Briggs box plenty of hope. I have started to develop the simplicity element of a CEO, but it's been a tough journey for me to go on, particularly starting as a scientist.

Milton Chang was a mentor who invested in my company Light Solutions and told me his secret weapon was the Enneagram, a framework for analysing personality that enabled him to quantify the personality traits that he could best work with in his investee CEOs and founders. Sir Peter Abeles, managing director and later board director of transport company TNT, used a similar tool to build what would become billed as 'the second biggest transport empire in the world'.[143] Several other Australian companies copied their 'empathy selling' technique, including co-founder of Zeetings and former Vice President and General Manager of Yahoo!'s commerce group Tony Surtees, while emotional intelligence coach Chris Golis has also taught many Australian companies this method of sales, captured in his book *Empathy Selling*. Entrepreneurs absolutely need this approach in their toolbox. You can tell so much about a customer by their desk, their way of talking, or how they act. This helps you to shift your pitch to language that resonates – some want just numbers, others want vision, and what works well with one will fail with another. It's the same for pitching to venture capitalists – always do your homework before you get in front of them and, once you're in front of them, still read your audience.

When I think about the real magic of scientists who have made the leap, I think of Matt Barrie at Freelancer.com, Dr Peter Farrell of ResMed, and Dr Zhengrong Shi of Suntech. They're all scientist–CEOs who fit the entrepreneur mould but didn't fit the CEO mould at all. As their companies grew, they grew with them. They've learned how to be ambidextrous – they've learned when to put themselves in the CEO mindset and simplify; but they haven't lost the entrepreneurial flair and somehow they managed to inhabit both sides of the coin.

Use diversity as the compass

While we're rethinking what it really takes to be a CEO, we have an opportunity to create a far more inclusive model than just considering what training a person has had. Diversity is the compass to guide us through the ambiguity of innovation. The opportunity here is more significant than in many other industries because, around the world, STEM and VC continue to be predominantly male, and generally culturally homogeneous.

I met 'Uncle' Frank Levinson, the co-founder of Finisar, in 1998 and, over the years, he kept helping me in various ways. I couldn't figure out why, because as much as there was a 'pay it forward' culture in the Valley, he was going above and beyond. One day I asked him. It turned out he had noticed the diversity of my hiring choices. He said:

> You've got this really interesting way of picking people that seems to have nothing to do with their gender, their age, their cultural background, or their religion. There's a particular way of thinking that people get stuck in, but you're not stuck. That's what's interested me about you because you think really differently. I think there is a magic in diversity.

When Frank was struggling to build Finisar, which would become a multi-billion dollar public company, he had an epiphany about the importance of empowering people to tell you when you're wrong. That was the reason that, after the IPO, Frank was still driving the same old car to work, still turning up at 7:30 in the morning and still working in the engineering lab with all of his team. He didn't think of himself as the boss – he thought of himself as an equal member of the team: maybe the first among equals, but absolutely equals, and so his team always felt empowered. Whatever ideas they had, whether it was a first-year engineer or a 20-year veteran, they all had equal value. That really helped Frank navigate his company through a lot of ups and downs and keep staff through difficult times:

> We had to do that in order to attract great people. [Finisar co-founder] Jerry [Rawls] had a vision of hiring smart people but they also had to be nice. Such people work well in small teams and small teams are a core part of our success. When we hired someone and they turned out to

have ego issues we would often disengage on that basis. Engineers like to win, so it was important for us to win and not just pay well. We could even keep people who were offered more money if we were perceived as empowering the engineers who were working for us to have their ideas enter the market and make the world better.[144]

Frank's inclusive culture was my first wake-up call, because I hadn't thought about it that much and I realised that while we might have had diversity of age and culture, the diversity on gender was really low.

It was so hard to find female engineers and female founders in the Valley, and, as a result, the whole culture had become male-dominated, creating a narrow way of thinking. My former fellow partner at Southern Cross Venture Partners, Tristen Langley, said it wasn't always easy being the only young woman in the boardroom:

I learned the hard way that you had to have thick skin. Roll with the punches. Don't whinge or whine. Just get the job done. You had to have a little competitiveness in you to handle the amount of testosterone around the boardroom table. Sonja Hoel from The Perkins Fund, and formerly with Menlo Ventures, had great insight, she said, 'Just be the best! Also, be quiet, listen and observe. The best remarks will be respected with great timing.' A key insight was you had to certainly establish technical credibility early on.[145]

Women in science are not as empowered as they should be, but I think they're more empowered in Australia than they are in the US, certainly more than they are in the Valley, and that gives Australia an opportunity to create a richer, more innovative system built on diversity and inclusion. Today, Tristen thinks Australia has an opportunity to wire gender diversity into our innovation system DNA because we have the benefit of forming it at a time of rapidly growing gender equity. She points to female CEOs like Melanie Perkins of Canva as blazing a trail for young Australian girls thinking about careers in tech, and because the total number of tech entrepreneurs in Australia is so small, women represent a much higher proportion than in the US. Now running her own venture and technology investment firm Amalfi Capital Management, Tristen said the number of female partners and co-founders of funds in her circle is 'quite impressive' and, in part, attributes the shift over her two decades of working in venture to the democratising nature of new platform technologies:

If you look at Meta or Facebook, 70 per cent of the people using Facebook are women, the same for Instagram. Even booksellers are about 50/50 men/women on Amazon. These are fundamental platform changes in the biggest companies in the world now and have levelled the playing field with access to women. So you have female-led technology companies cropping up and being successful because the ultimate customer is another woman. The predominant consumers online are women when it comes to ecommerce.

Tristen notes that, even outside of digital or internet startups, more women are leading deep-tech companies, although a gender divide remains between domains, saying: 'The healthcare sector and a deep-tech sector like semiconductors are night and day for women and men participation – but they are equally deep-tech.'

The ON program

CSIRO created the ON program (discussed more in Chapter 6) to accelerate ideas off lab benches and into startups, and scientists out of research jobs and into becoming founders. We didn't set it up with any diversity metrics, but women were successful because we redefined what we were looking for in a founder and a CEO, as well as removing some outdated barriers to participation, like scheduling that clashed with school drop-off and pick-up times. It recognised a different style of leadership and entrepreneurship. It realised that scientist-entrepreneurs look very different to all the other kinds of entrepreneurs and to all the other kinds of businesspeople. Being more open minded made the program far more naturally diverse, not because it was designed so that women would be more successful or biased in any way. Instead, it was designed to acknowledge that you're looking for people who think differently to really do innovation in spaces like science.

Unfortunately, some of our female founders still ran into trouble once they went out seeking investment, because the investors hadn't caught up (and some still haven't), but I hope we're getting there. Leonore Ryan was the founder of Cardihab, a secure telehealth app to connect cardiac patients with their doctors through their recovery from home. She had been part of the team to take the idea behind Cardihab through the ON program, where they discovered that 70 per cent of people who have heart attacks do not return for follow-up treatment or care. The team realised how many people's lives could be saved with their technology and became locked on course.

Going in to pitch, Leonore knew that the culture of the cardiac community was not going to respond well to her as a female CEO and when we went to raise the money from three cardiologists, that was their first and main point. They said they were not going to fund it because Leonore had no experience in the medical market, but that wasn't the real reason. She was selling into a market that is dominated by men and they thought her customers weren't going to respect her. Their excuse was lack of experience, but she'd grown her own experience and she was creating a new market, so nobody had experience in it other than her.

I had a lot of faith in her ability to be the CEO in this company. Before going through ON, Leonore had been a scientist and worked in commercialisation of science – but she'd never been a CEO before. She went out, got customers, and started working and getting momentum. When we had to bring in the funding, I met with the investors and said, 'Look, I really want her to have a shot at being a CEO. What if CSIRO funds the first year, you put your money in the company but we don't spend it and we'll fund it as a safety net and at least give her a shot?' But they were just so stubborn about it; they said, 'No, it's our money and we can't waste a year.'

Despite the doubts of the investors, Leonore stayed on as Chair and CEO until the company completed its capital raising, but at that point, as I've seen so many times before, she was replaced as CEO.

While I was frustrated by the situation, I was so proud of her, because founders should bleed for the company. The fact that she wanted the company to be as successful as possible and was willing to do that even if that meant she had to suffer made me even more stubborn in wanting her to be CEO. That's the characteristic you're looking for in a founder. It's one of my few regrets about the ON program that I couldn't find a way for her to actually be the CEO after it left ON.

It meant that when Dr Silvia Pfeiffer, our next CEO who came out of ON, was going to do telehealth company Coviu, there was no way that she was *not* going to be the CEO of that company. Fortunately, in that market, the investors were a bit more broadminded and it wasn't really an issue. Although, I suspect in the back of their minds they thought, like so many investors do, that a scientist CEO was probably not going to be around in the long term, and they'd have time to find someone else and make the change.

How wrong they were! Silvia has done a fabulous job of Coviu, and with the advent of COVID-19 she's seen phenomenal success. Silvia was able to go through that transition from 'I built this amazing technology' to true customer intimacy. I often think about how hard it must have been for her because her product's virtual, so she may never meet her customer. Yet she figured out how to get to know the customers and how to understand them through data and through their interactions with the platform. She also had the insight to connect with clinics who might want to host the platform and understand what they needed. She went along that journey of inventing whatever she wanted to inventing what the customer needs.

Leonore and Silvia are very different people, but both are absolutely obsessed with the mission of saving lives. They would both put their companies ahead of themselves, as a good founder should. They are missionaries who inspired followship in their teams. Followship is the critical outcome of leadership. The really great founders realise they may not be the best person to do everything in the company, so they value and empower the team. Some of them even have an epiphany where they realise that maybe they shouldn't be the CEO at some point. I would love to see more scientists go further down that journey, because the leadership of the future needs scientist CEOs and scientist-founders more than ever.

The gender lens was such an eye opener in ON, because if you speak the language of growth, and innovation, and possibility, and opportunity, and collaboration, it seems to naturally resonate with scientist-founders and clear the way for female founders. We still have barriers in changing investor perceptions. Australian venture capitalists and investors still look at a CEO through a very traditional lens, forgetting that entrepreneurs, not CEOs, run startups. Australia's got to grow up on that front and broaden what it's looking for.

VC is not just a male industry – it's a white male industry and that further narrows the definition of a CEO likely to get investment. In 1999, I co-founded AOC Technologies with Gordon Gu, a brilliant laser scientist who I'd known while we were both doing our PhDs. Gordon had been, unsuccessfully, trying to raise investment to start the company. He understood that as a Chinese scientist in America trying to raise money, the system would be biased against him. But between us, we realised that we could probably bootstrap it without needing to raise money.

In the telecom boom, we built a company in the cable industry, peripheral to telecom. It was against all conventional wisdom, as the industry was suffering from a lack of science and engineering talent because everyone wanted to get into telecom. No one wanted to fund cable, but that meant Gordon had almost no competitors. We built an amazing company, completely oblivious to the telecom bubble. It wasn't a multi-billion-dollar company, but it's still going today. It employs a bunch of people in California, where he lives, and a bunch of people in China, in his hometown of Wuhan where he built the manufacturing plant to give something back to that town. Wuhan was a sleepy little provincial town when we started AOC, but while AOC grew well, Wuhan's population exploded.

I realise now that I learned so much from my interactions with Gordon. Until I worked closely with someone from a different background, I would never have understood the type of biases and prejudices, even in a place with a large Chinese population like San Francisco, that he needed to deal with, particularly in the VC space. It's better today, but back then, Gordon probably didn't have a hope of raising money and yet he had such a brilliant idea and was able to build this business based on great science and technology. He didn't fit the mould, he didn't fit the pattern, and yet he was able to do great things.

I'd met another great Chinese scientist, Xiaofan (Simon) Cao, in early 1989 when I took the train from Kowloon to Canton hoping to get a glimpse of China, but I didn't know that the station was on the People's Republic of China's side of the border and, not having a visa for China, I would have got far more than a glimpse. Simon was a young student who I got talking to on the long train ride, and he realised my mistake and got me back onto the return train without incident. Simon went to the US to do a PhD at UCLA and, along with each of his four classmates from countries outside the US, founded US$5 billion tech companies including Etek, Oplink and Avanex. It was one of the greatest examples I've ever seen of the power of diversity and the scientist founder. Simon was CTO of Avanex, which led the 'telecom wave' that preceded the creation of the internet. Simon made it onto the cover of *Forbes* and it was gratifying to see a Chinese American beat the odds, especially a scientist founder. Deregulation of telecommunications in the US created a bonanza of tech startups to reinvent the network from electrical to optical. It was also a bonanza for cultural diversity, because so many of the optical tech founders came from non-US backgrounds. It's a great study in the power of deregulation and market shift, and a glimpse of what we will see in the energy transition yet to come.

Dr Zhengrong Shi was a Chinese PhD student at University of New South Wales while I was still in Australia and studying at Macquarie University. Zhengrong and I became friends; we used to drink beer at a little pub in Kensington and he used to dream about starting companies. He wanted to change the world with solar. As I look back on it, he was a little bit like Elon Musk at Stanford, dreaming of space, electric cars and solar, and he had that dream from the beginning. Zhengrong was on a mission to do something remarkable in solar and he was convinced by a breakthrough that his University of New South Wales supervisor, Professor Martin Green, had made. At the time, everyone was focused on getting the best efficiency solar cell, but Martin was all about how to get the combination of cost and efficiency, because the optimum is the combination, not one or the other – a great example of collaboration versus competition.

Zhengrong spent a year trying to raise money in Australia to fund SunTech and no one would take him seriously: he was Chinese, so that wasn't going to work very well in a white male dominated VC industry; and he was a scientist, so what would he know about starting a company? No one could really see it here, but when he went back to China, he raised about US$10 million in about 2 months. Our narrow-mindedness was their gain and our loss. Today, we are still struggling to catch up to Chinese investment in renewables, despite much of that IP originating in Australia. Zhengrong gives so much back to Australia. As mentioned in Chapter 2 Zhengrong is now on the board of the next generation of Australia solar with Sun Drive, which is showing all the right signs of becoming a thriving job generator for Australia, supported by deep-tech VC from firms like Blackbird and Main Sequence.

A brighter future with scientist CEOs

We have to think differently about what characteristics a CEO should have, particularly when we're dealing with scientists. We have to see a different pattern. Diversity counts in every single dimension because commercialisation of science will never be a one-size-fits-all proposition. The artificial separation of careers and capabilities has to stop if we're going to navigate the transition into an innovation economy. We can't think outside of the box if we're stuck being squares. Most businesses need more science in their strategy and the future chief executives of Australian companies can't be from central casting.

I'm absolutely convinced by my experiences that Silicon Valley, maybe even the rest of the world, has missed a trick around the power of diversity and the paradigm shift needed to really see the unique pattern (and there is a pattern) for a scientist CEO. If we can get that right in this country, then I think we have a chance to compete remarkably well with the rest of the world in our commercialisation of science.

4

Pathways for scientist CEOs

In the previous chapter, I made the case for why Australia needs scientist CEOs; in this chapter I tackle how we can support our scientists to make that quantum leap in career. It would be naive to suggest there is no gap between the skillsets of a scientist founder and a deep-tech CEO, but I think the leap is smaller than most scientists – and probably most venture capitalists – actually think. Further, I would argue that when we equip our scientists to lead deep-tech companies, they are vastly more capable than traditional CEOs in leading these kinds of companies. This chapter suggests three important ways we can better set our scientists up for a successful transition to CEO: by giving them more exposure to business basics in their undergraduate years (or a crash course afterwards without packing them off to do an MBA); by steering them away from perfection in the lab and towards perfection in the marketplace; and by rewarding scientists who say their research will have one, specific, crystal-clear application in the world, instead of rewarding scientists who see a million ways their work will make the world a better place. These are not behaviours inherent in a scientist; they are the metrics we've decided to use to measure their success – and we have the power to change them.

Break down silos

After Iridex went public, we created a couple of startups within it and hired two very smart physicists I'd studied with at Macquarie University in Sydney: Dr Brad Renton to run our dermatology arm and Dr Dave Matthews to drive OEM (Original Equipment Manufacturer) laser sales. Brad learned the importance of market understanding and became a deep domain expert in medical lasers, while Dave pursued technology and science and has worked in some extraordinary cutting-edge laser companies – both sticking to the same field of lasers they completed their PhDs in decades earlier. Many scientists tend to think they must make the most of what they learned at university and rise to the lofty heights of that particular field. I surprised them when I left the laser field to dive into the medical device industry, then telecommunications, then semiconductors, then VC. Jumping out when you are near the top of a field to land at the bottom of another is risky, but it's how you learn and continue to innovate. While I'm confident of my past successes, I never try to lead in a new field based on my track record. I try to stay humble and always be a learner. That's the beauty of good VC – it's not about teaching entrepreneurs you invest in, instead it's about partnership and learning together as you navigate something that's never been done before. Risk aversion stops most people from doing this reinvention. One of my mentors, Milton Chang, said one of his secrets to great investing was to look for people who reinvent themselves about every 5–7 years. I want more scientists to see their degree as a lever, not a manacle – see past it.

Open universities

Scientists shouldn't be told from the day they enrol in a science degree that a business education is closed to them. In fact, their science experience may prepare them to run companies far better than business or management courses do, especially in the innovation revolution where none of the old rules apply. In Australian universities, we make it incredibly difficult for students to pick up subjects outside of their core qualification. When I was finishing my Physics PhD at Stanford, I could sit in on business classes and learn about how to be part of the Silicon Valley ecosystem outside the classroom door. But in Australia, our courses are much more formal. CEO of Freelancer.com and Escrow.com Matt Barrie taught at the University of Sydney from 2001 to 2014, first in cryptography and then entrepreneurship. He left out of frustration with not being able to open his classes to the students he felt needed it most – science students:

> It became too hard to lecture at university because it was just so bureaucratic. I was teaching a multi-disciplinary course like entrepreneurship, which, if you taught it at Stanford, you'd have people attending from medicine, engineering, business, from everywhere. I was doing it through electrical engineering and you could enrol from electrical engineering, but you couldn't enrol from computer science, even though it was the same faculty, because it was a different school and because the course codes were different. The mechanical engineers couldn't enrol. The aerospace engineers couldn't enrol. The chemical engineers couldn't enrol. The business people couldn't enrol. I took the class seriously. I had incredible guest lecturers like the Head of Data Science for Amazon globally, the Head of Product for Mobile for Facebook globally – in fact, the speaker from Facebook was actually one of my former students. Everyone who is anyone in tech in Australia

spoke to that class, everyone from Mike Cannon-Brookes to Ruslan Kogan. I was constantly frustrated because the class would only have 40 people in it. I would open up the seminars to everyone, and the lecturers would come in and say, 'Why aren't there 1000 people in the room?' I had these incredible speakers and every year, everyone complained they couldn't enrol for credit. I had people attending the class who weren't enrolled and who weren't even university students. They would come from industry and just sit in the class at the back because they liked it. It was very frustrating doing all that work and then not being able to broaden the access.[146]

ResMed founder and Chair Emeritus Dr Peter Farrell said going to MIT on a student fellowship 'was a complete eyeopener':

You had professors that were out starting their own businesses and I just thought, wow, they're getting stuff out into the marketplace. This is what it's all about. Of course, chemical engineering started at MIT in the late 1890s and being a student at MIT had a major influence with respect to the value of entrepreneurship (opportunity seeking as opposed to risk taking) and innovation (where someone writes a cheque). Innovation only occurs when a product or process is delivered into the marketplace and it solves a problem. No cheque? Then innovation doesn't exist.

I thought, well, if you're doing something, you're spending time and effort on it, you want to make it work as a business. MIT, and Stanford on the west coast, were doing similar things, and it became part of my DNA that if you're going to work on stuff, you've got to make sure it's practical and you're going to get something out of it or do something with it. You don't do it for money, the money flows if you get it right. In Australia, there wasn't that sort of entrepreneurial focus, but obviously the US has been doing it for decades.[147]

With qualifications in chemical and biomedical engineering, Peter completed his PhD on treatment using an artificial kidney, which is how he came to be hired by Baxter International, which had a billion-dollar business in that space at the time. Chris Lynch, who would go on to become Peter's partner in founding ResMed, initially declined Peter's offer to join him at Baxter as he 'didn't know anything about medicine', having an undergraduate degree in chemical engineering. Peter told him he didn't need to know anything about medicine – he just needed to understand rates of change, which Peter said was 'bread and butter for chemical engineers'. After Chris was given a few journal papers to read, he took the job, realising he understood the concepts well and he just needed to brush up on his medical lexicon, demonstrating the broad applicability of STEM skills.

Peter had started an MBA – numerous times – but never finished it because he was ultimately too busy, ironically, running businesses that have gone on to be case studies for MBA students:

> Stanford and Harvard have both done case studies on ResMed and I used to make the point - because the MBA students will tell you how you should go forward and make something really happen without actually knowing much - whenever I talk at a business school, I'd say, listen, entrepreneurship is not about risk taking, it's about seizing an opportunity. It's just that the guy who's doing that sees a little bit further over the horizon than the average Joe and backs his judgment, seizing an opportunity. People talk about innovation, but mostly they're talking about creativity. Innovation only occurs when somebody writes a cheque. End of story. If it doesn't get into the marketplace and somebody said this is going to solve a problem for me, and I'm going to write you a cheque, that's when the rubber hits the road. That's when you say, 'Oh, gosh, maybe we do have an innovation here.'[148]

Founder and CEO of Soprano Design Dr Richard Favero has a PhD in electrical engineering, and while he said the technology has moved on in his field, he still finds his PhD useful today:

> I'm not a big believer that education should be very vocational. You're studying because you want to learn how to learn. The processes that you go through are much more useful in the way you apply them afterwards, that's where the real power comes from in research degrees. Society is in a zone where everyone feels like they've got to study a course to do a job, rather than just study a field and then apply those skills in the field of whatever you ultimately choose to do. In a lot of ways, I found the rigour and the attention to detail in the program of the PhD much more useful than the content ever was. You'd be surprised how many times, particularly today, even though it's been 30 years since I finished it, that AI and big data, which is exactly what we were doing in that speech lab all those years ago, are relevant. You never know how it's going to be useful to you.[149]

After founding Soprano nearly three decades ago, Richard said scientists do need some business literacy to protect themselves:

> I've watched a lot of good tech guys get absolutely messed up and their equity completely diluted by very predatory behaviour by those who are very sharp. All engineering and science students should take a course on how equity is raised, how a convertible note works, what's a liquidation preference or preference shares. They're all smart enough to know because the maths isn't that complicated, but you have to realise how it actually plays out. Who teaches you that stuff? Regrettably, it's the advisors who are typically not on your side.

> One of the big things we miss out on at the moment is an equity-raising founding-business course - it only needs to be one semester course for two hours a week to actually enable scientists so that when they're sitting in front of any of the VC/PE [private equity] funds that they can actually say, 'I actually do know what a liquidation preference means and I don't think I'm happy with that. If you want to sit there equal with me then you should come on the ride for that valuation.'[150]

Experience over qualifications

When I was developing my first product, before I'd even created the company to commercialise it, I was obsessed about getting it in the hands of a real customer, but I had no money to hire engineers to help me. I learned CAD (computer-aided design) programming so I could design it and build it – including converting my basement into a lab and learning how to build the unique power supply and control electronics that my laser needed. I'm not great at electronics, but there were books that taught me how to build the unique power supply I needed to power my special laser. I needed to patent the idea, so I read other patents and figured out how to write and file my own. This behaviour isn't what I'd recommend today, but it's a great example of the focus you need to be an entrepreneur. If you go through the pain of doing these things yourself, you will forever value the people you ultimately hire to do them for you. Too many scientists are too narrowly focused within their specific fields, but it's amazing what you discover when you venture way outside your field to engineering, law, business, or any other field your startup needs to succeed. The attraction of VC, of course, is that you can pay others to do all this for you and it saves massively on time to market, but don't forget that every experience of adversity makes you stronger. Today physics is a contracting field, with fewer enrolments and fewer physics major degrees available as job growth isn't in 'physics' but in engineering or computer science or coding. I suspect it's declining the same way degrees in geology have declined with fewer jobs available in oil and gas. Maybe I should have done engineering – but physics teaches you how to figure out how everything works and not to be afraid to change it.

My scientist co-founder at Lightbit, Dr Rob Batchko, had the potential to be a CEO, but he thought he'd have to act the part of the 'businessman' to achieve that. He read up on business books and tried to 'talk the talk', but it made him sound inauthentic. He hadn't lived in that world, so it undermined his credibility. He would have been better off surprising people with what he did know about business rather than projecting business expertise. Interestingly, after Lightbit, Rob founded his own successful company, focusing on leading with his natural strengths.

Ultimately, you need to be ready to roll your sleeves up and 'carry a bag' like a real salesperson. Everyone needs to be a salesperson, but especially the CEO. A good salesperson will tell you it's all about charisma, charm and the pitch, but a great salesperson knows it's about listening not talking, but when you do talk, it's about authenticity and trust. Lead with what you know, then work with your venture capitalists on the rest, because people love to teach and share their knowledge. Good investors should be your partners.

As a venture capitalist, I was only interested in funding founders who I could really help by filling gaps in their skills (otherwise what right did I have to invest in them?). One of my first experiences with this was when Intel wanted me to be the CEO and take over a company called Intersymbol. Intersymbol was a spin out from the University of Illinois Urbana-Champaign, founded by Dr Andy Singer who had been the protégé of Nobel Prize winner Jack Kilby, who had won the prize for his work with semiconductors. Andy was also an absolutely off-the-charts genius at semiconductors and digital and chips. He'd come up with a way to counteract noise distortion on signals, a very clever and unique idea that was basically forward error correction. Like most scientists, he led with the brilliant invention and secured corporate VC through Intel, which is between corporate investment and traditional Silicon Valley, because Intel is the size and age of a big corporate but has its origins and DNA in the Valley.

Because Andy was a scientist CEO, he was hard for other corporate investors to understand. Intel invested because the leadership could see the value of the tech and they knew the market, but they also thought they couldn't really trust Andy to run the company, because while he knew the science, he didn't know the market. They did what any corporate venture capitalist would do: they tried to find a CEO and they came after me. I flew out to spend a week with Andy and his team. I was convinced that he could actually be a CEO and we had a massive fight because he saw himself as a scientist, not a business leader. The fight took our relationship to a whole new level based on our deeper understanding of how each other saw the situation, and I agreed to be the executive chairman, but not the CEO. I committed to Andy that I would work with him to help him become the CEO as he went through the school of hard knocks, as we all need to do. That was fabulous experience for me, hopefully for him as well, because it really showed me how much I enjoyed nurturing and mentoring scientist CEOs and helping them be successful. We went through the telecom boom, the telecom bust and helping him sell his company.

Business degrees

Does a scientist need an MBA to run a business? Investors who don't understand science want to de-risk their investments by bringing in MBA expertise to run the company. Scientists who don't understand business can often fall prey to investors and other business leaders who take advantage of their naivety. I think the answer is somewhere in the middle – we need to make it easier for scientists to access the foundational tools to run a business, especially the tools needed to negotiate the early stage investment that will be essential to leaping across the Valley of Death and ensuring their role in the company isn't diluted in the process. If we want to build an economic pillar from science, we need to teach scientists how to be part of the economy.

Science-driven startups shouldn't be led by people with MBAs. MBAs tell you how to run traditional companies, and truly innovative startups are not traditional companies. Steve Blank said the key insight behind developing his Lean Launchpad method was realising that startups aren't small versions of large companies; they're doing something quite different; large companies execute business models while startups search for business models.[151] Blackbird co-founder Niki Scevak agrees, noting Steve's most important contribution through the development of the Lean Launchpad model was:

> delineating the difference between the process of discovery – going from 0 to 1, creating a company, discovering whether your product is worthwhile to a certain set of customers or not – and the process of optimisation and scaling, or going from 1 to 100. 100 per cent of our education is dedicated to the 1 to 100 phase or the optimisation of something that works. If you summarise an MBA, they assume that the thing is working and then they teach you about making it work to the best extent possible versus teaching you the scientific discovery process and the hypothesis testing that is the 0 to 1.[152]

The phrase was coined by co-founder of PayPal and Palantir Peter Thiel in his book *Zero to One*, where he describes the key insight: 'Doing what we already know how to do takes the world from 1 to *n*, adding more of something familiar.

But every time we create something new, we go from 0 to 1. The act of creation is singular, as is the moment of creation, and the result is something fresh and strange.'[153] If your company is truly innovative, there is no experience a CEO can bring that compares with the knowledge and passion of a founder who has developed their business from just an idea.

Most startups don't refer to business handbooks to solve problems, because they know doing things the way they've always been done isn't very innovative. If you look at Zappos or Netflix, you'll see they've all got unique cultures because their founders didn't really know how to build a business, so they learned new – and frankly better – ways of building businesses, because they were naturally innovative. Australian software company Atlassian is a great example of that. As it grew and became more successful, the partners from the Valley VC firm Accel, which had invested a lot in Atlassian, very carefully and politely suggested bringing in experts from the US to help co-CEOs Mike Cannon-Brookes and Scott Farquhar by putting other expertise around them. I can only imagine the number of arguments Mike would have had by flatly refusing their advice if it didn't make sense or seem right to him – a bit like US President John F Kennedy's criticism of the advice from older, 'wiser' people to invade the Bay of Pigs. Atlassian did many things that an MBA-trained CEO would have told Mike and Scott not to do and if they had followed the MBA advice, the company probably wouldn't have been as successful. They didn't know what they were doing in running a business because they'd never done it before and they didn't go to business school, but they just did what made sense to them and it worked exceptionally well. Scott said:

> I really think if we had grown up in Silicon Valley and had venture capital around us, we would have built a very different company that wouldn't have been as disruptive as ours has been. Because we didn't have venture capital, we didn't have people that had done it before to drag us back to the mean. We just grew up without anyone telling us the way it couldn't be done. Our experience of computers was downloading and using computer games.[154]

One of the few scientists I know with an MBA said his qualification didn't help him run a company, but it did open doors for him to learn how. Dr Josh Makower has founded multiple successful medical device startups, and is a Special Partner at US healthcare venture firm NEA and a Professor at Stanford, where he co-founded the School of Biodesign, inspired by his own background combining engineering, medicine and business. He said he got his MBA after his science qualifications got him a job at Pfizer, but the science background alone didn't give him the lexicon (as Dr Peter Farrell said) that he needed to get the job he wanted:

> My first job out of medical school was as a technology analyst at Pfizer. They needed somebody with a medical, clinical and technical background to help them with business deals that involved a variety of medical technologies. For the most part, my role was to chase down due diligence to evaluate technologies and understand their clinical applications. I was always frustrated, however, because, as soon as it would come to evaluating the product or company's financial value, they would say, okay, you can leave the room now, we don't need your input anymore. That's why I went to business school. I wanted to get a basic education in business to understand financial language and these concepts and I did not want to be viewed as just a technical person. I found that the business concepts were pretty easy to understand, but the ability to speak the language of business was also important. Having an MBA degree got me into the room where it happens.
>
> I think a scientist can be a good CEO, but it's a completely different skillset than being a technical expert and you need to respect the value of leadership and management. Our limb-saving company Transvascular was my first opportunity to be a CEO. I was having some significant issues managing the team, so I stepped over to Chief Technical Officer and we brought in Wick Goodspeed as CEO. Thankfully, I stayed on board as the CTO and I got a chance to watch how Wick Goodspeed just cut through and solved all the problems that I was losing sleep and developing an ulcer over. I realised there really is something to management experience, and I also realised that I didn't have it. I've stepped in as the CEO temporarily in other companies since then and I feel like I can do it now, but only after having learned from watching and working with great leaders. Before I had that experience, I struggled and it was … not pretty. Having an MBA certainly does not teach you how to run a company.[155]

People with MBAs are smart people, but assuming this means they can run a science startup is a greater mistake than thinking a scientist can't. Spreets founder and CEO Dean McEvoy thinks 'entrepreneurialism is like a virus you have to catch, rather than a course you have to teach', but I'm not sure we've quite bottled that virus yet.

Embrace near-death experiences

Throughout this book, I talk about Australia's fear of failure and the chilling effect this has on innovation. The scientific method doesn't see failure as a disaster – it merely disproves a hypothesis and begins the search for the next one to test. Australian scientist CEOs can create a new model for leadership in our country that embraces professional near-death experiences – those moments when failure seems inevitable – as not a failure of leadership, but rather an epiphany that can lead to the stronger return of the company. Many leaders are also driven by actual near-death experiences. Part of what drives me through technical challenges is remembering that I've survived being caught in a rip at the beach as a kid and nearly falling off North Head, part of the headlands at the entrance to Sydney Harbour, and that no technical challenge is as hard as the impact of my father's death when I was young.

In the Valley, deep-tech venture capitalists know that failure points are inevitable in startups, but they're rarely showstoppers or the end of a business. When I was running Lightbit, our first near-death experience was a failure to see our product would not easily fit into the existing telecommunications paradigm, discussed later in this book. We discovered the technology wouldn't work for telecommunications as you needed to heat the optical chip to a temperature way above the Telecordia spec. Worse, it turned out this was a known problem by the inventors. When we realised this, I tried to give the money back to our two top-tier venture capitalists, Accel and Mayfield. The venture capitalists were bemused – when you fail as a startup, good venture capitalists don't expect you to throw in the towel, they support you to dig in harder. But that simple act earned great trust, so they gave us more funding to crack the problem, which would go on to make the technology even more valuable, and they continued to invest in us because of our honesty, tenacity and success.

We took Lightbit back to the drawing board to assess what the core innovation of this chip was and think through other applications. We ultimately repositioned it as an optical processor – something that could do with light what the silicon chip was doing with electronics – and found another application within the telecommunications sector. This time, we thought it could be a replacement for the repeater huts that were built every 80 km or so across the US to regenerate the signals travelling in fibres across the country. We took the tech to Bell Labs to help us develop it using their insights as a tech provider to AT&T, the largest telecom in the US. In 1999, Bell unveiled it at the biggest conference in lasers in the world, the Optical Fibre Conference (OFC), showing how it worked over the breadth of the US with no distortion in the signal.

That's when Lightbit was declared a billion-dollar company. Many companies had been acquired in optical switching before that for multiple billions, so it was a very reasonable valuation considering the market at the time. Unfortunately, we were just starting acquisition discussions when both the telecom and the dotcom bubbles burst. In hindsight, there was a terrible moment just before the bust where we could have sold it for maybe $800–900 million, but our investors quite rightly decided to hold out for $1 billion given the valuation the company had received. That offer never came. The market plummeted, and while the offers kept coming in, they didn't go up. We'd wasted time at the beginning of the company's life by chasing big telecommunications carriers like AT&T with the product they thought they wanted. By the time we started working with Bell Labs and Nortel to ideate, the peak of the market had passed and we'd missed our window of opportunity.

Fortunately, we had a Plan B. Whether it's the market, an earthquake or a terrible flood, it's always your call on how to respond, so as a CEO, you must always have a back-up plan. As an entrepreneur you must never give up; you either win or die trying – your investors will worry about when to give up, so you mustn't. Plan B was in skunkworks, or unofficial development, from before we were christened a 'unicorn' at OFC in 2000; in fact, it was a blast from my past. Kevin Kalkhoven had taken his telecommunications company JDS Uniphase to phenomenal success – it was the most valuable company in the US stock market for a brief flash of time. Kevin had tried to buy Iridex after its IPO in 1996 and personally invested in Lightbit, and so helped us connect with JDS Uniphase, as it was desperate for a blue laser. We used the optical chip to convert its telecom-grade laser diode into a high-power blue laser, a unique offering for the market, so the deal was approved pretty easily, despite the market crashing around us. We were the favoured child again. But then, just before the deal was inked, the JDS Uniphase Board brought in a new CEO who blocked all mergers and acquisitions. Did I mention you also need a Plan C? We sold it for a 'top quartile return' in the worst tech crash in history – but that's not saying much.

Although Lightbit had been a rollercoaster, as I looked back on the startups I'd known through my career, I started to see a pattern of 'near-death experiences' that made me try to 'join the dots', as Steve Jobs used to say, and to understand the pattern. Each time my startups thought we had it nailed, the market would throw us a curve ball. I realised these near-death experiences hadn't, in the end, been the problem; they had actually led to the solution. All six of my startups almost failed, but each near-death experience turned into the moment of the epiphany. The failure was the turning point for a profound insight and a shift in thinking that ultimately led to success. You can't look at these near-death experiences as nature's way of telling you to go find another job – see them for the great teacher they really are and redouble your efforts to win.

Co-founder of Finisar Frank Levinson recalls a near-death experience for his company that has remained imprinted in his memory nearly 30 years later:

> Our original transceivers used CD-style lasers from the largest manufacturer of this type of laser. We received batch #9546 (1995 year, week 46) which had some bad or fragile parts. We shipped thousands of devices and they went off like firecrackers in our customers' systems once they were in the field. We had two big customers – NEC in Japan and NewBridge in Canada – and we almost killed both companies. We took back more than one-third of our shipped devices that year and nearly died. But on 1 January 1997, we moved from CD lasers to VCSELs a year ahead of anyone else. They were a thousand times or more reliable and we saved the company; we were agile and willing to qualify quickly in order to make this possible.[156]

Chair of multi-billion-dollar iron ore company Fortescue Metals Group, Dr Andrew Forrest, said his career has been marked by actual near-death experiences, each a reminder to try new things every day. His childhood experience of being the first on the scene with his father of a horrific bulldozer accident in outback Western Australia and holding the victim's hand for the 2-hour drive to the nearest hospital gave him 'a serious appreciation for life and commitment to not wasting it. If we simply do things the way we've always done them, we will be wasting our lives',[157] which he said plays out in the culture of risk-taking and innovation at Fortescue. In more recent years, a near-death accident while hiking in 2015 put him on a long road to rehabilitation, creating the opportunity to realise his long-time ambition to study marine ecosystems at university, and ultimately gave more momentum to environmentally focused initiatives like Fortescue Future Industries, which was formed in 2018 and invests in renewable energies.[158] He said of the lessons he learned from professional near-death experiences in Anaconda Nickel, which led to its merger with Glencore Steel: 'It was a rough road for Anaconda, mixed with massive success as evident in its enormous productive capacity today and its strong future. Of course, Australian culture and media focus on the failure, but it wasn't the aspects of success my team learnt the most from, it was the failures – and that's what led to the creation of Fortescue.'[159]

Dean McEvoy's best-known success, Spreets, was born out of a series of professional near-death experiences, including both him and his first company, Booking Angel, running out of cash, and from an epiphany that made him change his business:

> I was this stubborn entrepreneur. People would say no to me all the time and I was like, yeah, whatever, they don't know what they're talking

about. I pitched Mike Moritz from Sequoia and some amazing VCs like Chris Sacca and all these people were giving me feedback and I was probably just not really listening. One day I had a conversation with the founder of Eventbrite, Kevin Hartz, a lovely guy. He said, 'You're a smart guy, you should be doing way more than this. Think for a moment about why your business isn't successful.' I told him about our key insight to drive traffic to new customers' websites from our existing customers' websites, and he said, 'That's really interesting. It's kind of like a business called Groupon in the US.' I kind of A/B-tested my way to the Groupon group-buying model without having heard of it.

Soon after that, my grandfather died and I flew back to Sydney, completely broke. I couldn't face getting an average job, so I asked Phil [Morle] and Mick [Liubinskas] from Pollenizer for some help to try out this new model. I had no team, I'd exhausted all my other funds, and they agreed to help out with resourcing in return for some equity. We built the first version of the group buying site we started, which was called Spreets, and we launched it on my birthday, 4th February 2010 and sold it to Yahoo 11 months later.

I originally tried to pitch Spreets as part of Booking Angel and no one wanted to talk to me. They were like, Dean, give up on that thing. So psychologically and emotionally I had to kill the baby and then start a new company and recapitalise it and get some money in.[160]

Yahoo! acquired Spreets for around $40 million at the end of 2010,[161] one of the biggest Australian tech deals at that time.[162] In a testament to the mores of missionary deep-tech CEOs, Dean paid back the investors in his earlier company, Booking Angel, out of his Spreets gains because 'they backed me in the learning process that brought me to Spreets, so it felt like the right thing to do'.[163] I did the same when we merged Light Solutions and Iris, repaying our original investors but letting them keep their stock in the new company, because this is what missionary CEOs do.

Near-death experiences are harrowing, but they are what separates the proverbial wheat from the chaff, because it takes a leader and a team with true belief in what they're doing to push through that adversity to create a stronger company.

Prohibit perfectionism

One of the reasons scientists become CTOs or chief scientists instead of CEOs is a passion for perfection and endless iteration. If a startup is truly unique, it can get away with taking a bit more time to get to market while inventing the product. But scientists have to be wary of inevitably wanting to make inventions perfect and spending too much time innovating and iterating in the lab when they should be in the market, learning from customers. When I was in venture in the Valley, I wrote a regular column providing advice to would-be deep-tech founders. I met many passionate scientists who had brilliant ideas, but I also saw many of them fall into the trap of believing starting a business was a great way to carve out time and space for their passion projects. In May 2001, I wrote:

> Don't get into business to do research – find a university and give them some money to do it for you; they'll do a better job for less money ... As R&D people we learn there is no such thing as failure; even a null result is valuable. Not in business. If you spend a year working on a contract that then goes south, you just wasted a year. You failed to generate revenue and you took food out of the mouths of your team. You should be shot! I hope you had a backup plan.[164]

If I had known that many years later, I would be heading up one of those research organisations, I might have chosen my words differently, but the sentiment is still right. Passion for invention is a wonderful trait in a scientist but can be a liability in a scientist CEO.

In his 2006 memoir, *iWoz*, Apple co-founder Steve Wozniak reflected on having had a similar hope for the company:

> When we first started Apple, Steve [Jobs] and I really had this engineering-centric model in mind ... but we knew what we were getting into because Mike Markkula told us ... the product is going to be driven by demands that the marketing department finds in customers. This is just the opposite of a place where engineers just build whatever they love and marketing comes up with ways to market them. I knew this was going to be a challenge for me.[165]

One of Apple's contemporaries in the Valley, Osborne Computers, took this lesson too far. Osborne was so good at innovating that as soon as it released a new product, it'd begin talking up the next one. Unfortunately, that meant no one ever wanted to buy its current product, instead deciding to wait for the next one. As a result, it earned the dubious honour of creating the 'Osborne effect'.[166]

David Thodey saw this in action when he was CSIRO's Chair:

> I appreciate the wonderful intellect and analytical capabilities of our scientists, who are trained to look for different options and to test assumptions, validate research, and empirically use data to question what they are researching or testing, looking for optionality, which I applaud. I think all leaders need to be open to new ideas and new ways of looking at things. We should not be caught in complexity or optionality, but through exercising good judgment based on empirical data or on experience to get to the 'least wrong' possible solution. Maybe there is this inherent desire in a scientist to define the world in very black and

white terms, or empirical truth, when the reality is we live in an imperfect world. However, we should always keep searching for greater insights.[167]

Fortescue Metals Group Chair Dr Andrew Forrest not only saw this in the scientists he recruited, but he nearly fell prey to it himself during his recent PhD studies, when he became enthralled with understanding the potential of 'whole cell analysis' in marine biology to tell us about our oceans. He recalls:

> I'd frustrated my supervisor so much at this stage that she said, 'Well, if you want to now take on global warming and plastic pollution,' – which I did, I wanted to go and do a paper on what I called the 'necklace of death', which is the permafrost around the Arctic, and when I introduced lighting up the ocean to understand what is actually in it for the first time through whole-cell, full water column sampling and analysis, it was too much. She said, 'You're just never going to graduate. You're just going to be one of these professional students who just hang around universities. They all just get old.' In the end, we compromised and published five papers instead of the required three and I still graduated.[168]

He realised he preferred to be doing something about solving the problems rather than endlessly studying them.

Translucent

In 1999, I learned the Osborne lesson the hard way when I co-founded a company called Translucent with an Australian scientist at Stanford, Dr Petar Atanackovic. Petar's research had found a way to change the bandgap in silicon, creating many new ways to use silicon. I was so excited by his research and he was so excited by my ideas for applications. It never occurred to Petar that this breakthrough might actually be useful; he had just wanted to see if it was possible. In fact, this was a classic platform technology with a thousand applications, but we had to figure out the one that would work for us. The bandgap of a semiconductor fundamentally changes its behaviour so you could make a revolutionary new computer chip or possibly a quantum processor or you could solve a problem in the current semiconductor industry roadmap to enable the next generation of chip. How do you pick? It's that magical combination of a market opportunity, a seemingly impossible problem, and the ability to make something work within the first round of VC funding. Petar and I spent a year figuring out where we would bet our technology to work best and I wrote several patents with him. The internet was growing at pace, but bandwidth was limiting what could be done, so finding bottlenecks to unblock was a great opportunity.

The internet is enabled by lasers and optical technology, moving literally at the speed of light, but everything you interact with as a human, either on the phone in your hand or the laptop at your fingertips, starts as an electrical signal – even including the nerve impulses from your brain that trigger your fingers to type. There was no 'integrated circuit' or solution to turn electrical into optical and have them speak to each other; optics uses a plethora of different materials to make it work while electronics harnesses the power of just one material, silicon. We decided to try cracking the impossible problem of uniting optics and electronics into a common chip using silicon by making a silicon laser. Any second-year university physics student can tell you that silicon has an 'indirect bandgap structure' so it can't support a 'population inversion'; basically, it can't be used in a laser. But we thought Petar could change that with his invention – we hoped.

It was a crazy idea and almost impossible to pull off, but that's what appealed to us. Once we had demonstrated optical gain in silicon, we started attracting serious funding from a bunch of investors who were really interested in the company, including Intel. Next, we had to build an actual device, a minimum viable product, that we could get into the hands of a customer to start iterating it. But because Petar was so brilliant, everything he did was a breakthrough. I couldn't stop Petar from inventing and I didn't want to because he was a wellspring of genius. But ultimately we couldn't cope with it all because there were so many inventions and he really wanted to be involved in every one because they were his babies. After working with Petar for a few years, I realised that we were probably never going to be able to produce a product because we could never get him to let go long enough to commercialise it. At the same time, he created so much value through the IP the company created that we had an opportunity to sell it to Intel. Many would say we should have, because it would have delivered the highest total value and grown the market the most, but I saw a different way forward.

In the end, Translucent had a fatal flaw in that, while Petar had figured out how to change a fundamental property of silicon so it could lase, the way he did it was unmanufacturable. I told Intel this, because it was the truth and because I thought instead of selling we could partner to solve the problem. I told Intel that only it had the expertise to figure out Translucent's impossible manufacturing conundrum. What would otherwise have been an adversarial negotiation and due diligence – what we had done was hard to believe, after all – instead turned into a genuine partnership. Then something magical happened. Intel started ascribing value to Translucent that it did not have. Intel saw the true potential through its manufacturing expertise and projected the much bigger prize we saw, and it was willing to pay us some of that: the value of a deep-tech CEO who could see applications and implications for technology before the market did, as discussed in the previous chapter. It was a great lesson in negotiating the sale of your company: if you can help your acquirer to see the much bigger prize and be fair enough to give you part of that, it's a great deal. Intel and Infinera did ultimately crack the tech 7 years later, the latter delivering a NASDAQ IPO in June 2007.[169]

I didn't want to sell Translucent to Intel because I hoped to move Translucent from the Valley to Australia – which I'd hoped, in turn, would encourage investment in something to leapfrog semiconductor technology (successive New South Wales governments have been fascinated with the idea to build a semiconductor fabrication plant, also called a 'semi fab', in Australia for 20 years). I've always wanted to see an advanced chip manufacturing plant in Australia because such plants were the cornerstone companies in the Valley and Israel (as discussed in Chapter 7). I found a visionary CEO running an Australian public company in Dr Mike Goldsworthy of Silex Systems, who bought Translucent. Mike got it; he saw the potential. Mike's company was trying to do something equally impossible – use a laser to enrich uranium and give Australia's raw commodity a unique differentiator to capture more value domestically. Mike was a rare scientist CEO with a PhD in nuclear physics and I knew we had found a rare kindred spirit in Australia. It worked out great for Mike because his share price and market capitalisation slowly went up almost tenfold over the next 12 months, and he broke into solar and other aspects of semiconductors as well. It worked out great for Petar because he got to move back to Australia and now works with former Intel Fellow Dr Steven Duvall on next generation semiconductor fabrication, which he hopes will help Australia catch up on semi and quantum.

When I think back, we had so many options for Translucent's technology. If we had been successful in creating the product and built a public company, Petar would have spent the next 20 years inventing every product for that company. He would have been what Steve Wozniak was to Apple, coming up with breakthrough after breakthrough after breakthrough. But ultimately, his passion for holding onto his baby meant letting go of the plan to build a company.

This desire to continue iterating and perfecting is very natural in scientists. In university, we're taught to think that way about our IP and our research, but unfortunately no one can do it alone. No matter how brilliant they are, you've got to let other people in and gather them around you to form a team to actually take the product all the way to market. Successful scientist CEOs understand the value in partnering, letting other people get their hands on their idea so they can help. They realise that the brilliance of invention is about 1 per cent of the actual value in the company – the other 99 per cent comes from the people who deliver it. You have to value that more than you value your genius. Bringing inventors to the table recognises their important voice and sets an example for the next generation of aspiring scientist CEOs.

Focus on one, see the whole
Focus on one application

While entrepreneurial flexibility and adaptability is essential in business today, it can be a liability too early in the development of a company and a trap many scientists fall into, led by their missionary faith in their invention. Scientists most frequently trip up while building their company when they think their technology can do everything, because they see so many possible futures at once. Of course, their technology may well be able to do everything. But the success of startups relies absolutely on picking just one thing. One product. One application. One customer. At the end of scientific journal articles now, scientists are being increasingly asked to consider the applications of their findings, so they'll tell you it has applications in defence, agriculture, manufacturing, health and so on. They're all very general applications and there's a laundry list of them, rather than just one. When I developed my green laser, I could tell you it was good for ophthalmologists and dermatologists and hair removal and 3D-printing and drilling holes in turbine blades and photograph printing – but that was a recipe for failure.

Scientists struggle with this notion of myopic, blinkered focus on doing only one thing better than everybody else. But that's how you create your beachhead and the firm foundation a successful company needs. In a way, it's ironic because when you start your science training, you know a little bit about a lot of things and as you go along in your education, you know more and more about fewer and fewer things, until you are an expert in one very narrow field. Markets require the same precision focus – and then you need to go from zero to 100 at warp speed to have the same level of expertise about your customers. Both science and markets become very specialised very quickly, so you need to be an endlessly curious person who isn't afraid to ask questions. That's definitely at the heart of science training and, I think, at the heart of the best business training. When you're building your company, investors look for that ability to say no to everything else and just do the one idea. Investors want to see that, because if you believe so much that you're going to bet the farm, then they will too. They know that if you do that really well, that's your springboard to go and do other things.

Scientists need to think of customers as experiments and learn everything they possibly can from the customers who trust them and are willing to let the scientist fail with them and learn from them. Then the scientists will be able to go out and find another customer and another customer, in the same beachhead, in the same market, in the same application. You will know you are getting it right when you can start to talk about your company through the eyes of your customers – later when you write your pitch, strategy and plan for VC investment (there's an example at the end of this part of the book), it will all be through the eyes of a customer instead of a scientist, then you will have snatched the pebble from my hand, grasshopper.

Light Solutions

We formed Light Solutions in the early 1990s to commercialise a solid-state green laser in a chip, something that had never been done before. As a scientist, I thought the world would come knocking and, in a way, it did. We had plenty of customers. Polaroid wanted one to print photos; a stereolithography company wanted one to try some crazy idea called 3D printing; and Boeing wanted one to drill holes in turbine blades. We had 20 companies with 20 completely different markets and lots of different applications. The problem was we were making money building one-off prototypes for these unique applications and learning a lot as we went, but we couldn't consolidate our focus and start scaling up while we were so broadly committed. We had this plethora of opportunities, but we couldn't make heads or tails of them. So I went back to fundamentals and common sense – something I was still learning at the time – and looked for an application for this laser that couldn't be done any other way. Drilling holes in Boeing's turbines or supporting 3D printing couldn't be done by anything but a laser, but eventually they would be done by other lasers in the market, and I knew I didn't want to be a 'me too' company. I wanted to be unique, and I truly believed this laser could be.

I've always been intrigued by medical applications; I want to feel like I'm doing something to change the world and feel good about what I do. I can feel good about drilling holes in turbines if it's going to make the planes safer, but if I'm just trying to improve the fuel efficiency for the airline so it can make more money, I'm less interested in that. Eventually I came across Iris Medical, a company making a laser to cure blindness in diabetics. It didn't have unique technology, but it tried to jump on the solid-state laser revolution. It was selling infrared laser diodes to ophthalmologists as an improvement on the green gas lasers that were already in use, leveraging the convenience of a smaller infrared diode over a big, bulky gas laser installation that required high voltage, water flow and other unwieldy infrastructures. When I started talking to doctors, they told me the infrared laser diode was over-treating and burning patients because it took a lot more power to leave a visible mark on their retina compared to the green laser. They could learn a completely new way to treat using infrared, but that would mean unlearning years of experience with green lasers and they weren't willing to do that.

When Light Solutions merged with Iris Medical to form Iridex, we combined Light Solutions' new green laser technology with Iris Medical's small box infrastructure and solved the patient burning problem. Over time, we also invented the way to alter the pulse format of the laser (something gas lasers couldn't do) for different applications based on feedback from doctors. Now we're not talking about making a 'me too' product – we're talking about a completely different footprint, which is much easier to install and operate, and which has a much wider spread, a bigger market and unique features that the existing one doesn't have. Now we've got an innovation.

Redfern Integrated Optics

Around 2010, I became involved with Redfern Integrated Optics (RIO), a company spun out of the photonics group at the University of Sydney in 2000, in the peak of the telecom bubble. I started as an investor in RIO, then I became the Chair after it got into trouble, before finally becoming the CEO. Rather than just building one company, RIO thought it had such amazing technology that it created a group and a fund, including creating seven companies to support all the profound and broad applications of its technology. A number of local investors piled into this on the back of the dot com boom and while RIO did have amazing technology, it let the interest of the investors fuel their plans to create seven companies.

RIO had come up with a way to make a laser diode run at a perfect single frequency, which did have all kinds of applications because it's very hard to make a laser stick on a very precise frequency. RIO was in every market. It was chasing wind sensing for renewable energy because of the renewable energy boom; it was chasing communications because of the telecom boom; it was chasing data centres for the move to the cloud. The situation was extraordinary. In part, it did the right thing by looking for the big shifts, but like most scientists, the ones in RIO started with the scientific breakthrough and then went searching for a problem to solve with it. But the problem with really big shifts like renewables or telecom is that everyone can see them, so there is a lot of competition.

When I took over as CEO of RIO, I really narrowed its focus on to the wind market. We'd deployed this laser technology to sense the wind velocity approaching turbines, so you could tilt the blades to capture the maximum energy when the wind is low and if the wind picks up, you can tilt the blades so that you don't damage the turbine. I saw that as a massive opportunity – and I was wrong. What happened was what so often happens in tech markets: we had a massive wave of early adopters, which tricked us because we saw massive growth and a unique opportunity. We struggled. We sold lots of product and it got really exciting and got the investors really excited, but then the growth began to peter out. This is the third Valley of Death (see Figure 3, page xii), the most common challenge faced by startups trying to secure a sustainable customer base. To cross this Valley, you must evolve from technology leadership to customer understanding.

Other people used us to prove that the market works, but the market itself didn't value the technology the way we did. They figured out another way to do this wind measurement that didn't need this laser, which meant they no longer needed us. When your unique technology is adding value to someone else's product, as our laser was in the turbines, the makers of that product will look for ways to reduce that value and have other options because it's not good business to be dependent entirely on one person.

Eventually I recognised that I had made the wrong decision, but this realisation is really important because as scientists we don't give up on our hypothesis easily. We're stubborn and we keep going and we're not afraid of little failures, because we see them as a step closer towards proving our hypothesis. But we do get stuck on a hypothesis, because we want to prove that it was right. If you could harness that characteristic of a scientist to focus on the hypothesis and make them focus on the market, or a product, that would be a strength. But for some reason as scientists, we focus on that initial idea and we are not willing to let it go. It's ironic because, in our brains, we're thinking about all these other applications, but once we're in one, we don't really seem to want to pivot. It's a bit of a paradox. So you'd better be sure that you are unique in your ability and there's no other way to do it, because if there is, others will find it and they'll change their approach to achieve the same results more cheaply, and that will reduce the value of your company.

The initial revenue from the wind opportunity meant we could test some other markets. We discovered we could use the laser to monitor a border perimeter and detect footsteps at high levels of sensitivity, so you could tell if it was a bird or a dog or a cat, or a person, walking near your perimeter. This was especially valuable to security systems providers. RIO worked with a UK company called QinetiQ to develop a unique system, ultimately selling QinetiQ the company so we could get out, and turning RIO into an exit. RIO missed its chance to be a standalone company, but it did find an opportunity to exit, showing that when the technology is unique, you almost always end up with something of value to sell. Interestingly, that Australian invention found its way into the global market and back to Australia, although Australian customers wouldn't buy it until it was owned by the international QinetiQ.

While it was wonderful for Australia that there was a willingness among investors to back Redfern's efforts to build these seven companies and create all this activity, it also set us back because of the inevitable failure. Scientists will eventually become comfortable with failure because even a null hypothesis gives valuable information, but we have a long way to go in recognising a model of a CEO who is comfortable with it.

See the whole market

It's just as important to focus on one thing for your technology to do as it is to appreciate what other technology could do the same thing. Around 2014, then-Prime Minister Malcolm Turnbull asked then-Australian Chief Scientist Dr Alan Finkel and me to spend some time with Professor Michelle Simmons to understand her silicon quantum computing company before the government invested $26 million into it, which it ultimately did in December 2015.[170] Michelle had a great vision for silicon, similar to Petar's learning that if you can do it in silicon it will always win because of the success, scale and history of the semiconductor market. Michelle's competitors were going down the analogous path that the optics industry had done in the internet revolution, using dozens of exotic materials that probably couldn't be mass manufactured the way silicon could. Michelle's challenge was to reinvent silicon – bend it to her will – which was thought to be impossible, but I'd seen that movie before with Petar. I was surprised how focused she was on her quantum competitors but how little she worried about other ways to achieve the outcome of quantum without needing quantum, like semiconductor company Wave (mentioned in Chapter 3).

The market will change the moment you are funded – if Doc and Marty had gone back in time and funded a startup in *Back to the Future*, they could indeed have unravelled the whole space–time continuum. This is why agility is so important: you need to be focused on creating your one killer product, but you can't ignore the future as it changes around you. The moment people know what you are doing, competitors emerge to take your future from you. The incumbent players won't want anything to change and they'll try to stifle your innovation. Does this mean you should stay in stealth and not tell anyone? They tried this in the telecom revolution, where startups were all in 'stealth mode' until their miracle was ready to be unleashed on the world – but many were so secret they were also hiding from their customers. You can't pivot if you don't get feedback, so it's a balance. The future isn't fixed, and everything you, your competitors and your customers do will change your chances of success. So frequently revisit your vision; be honest with yourself about where you really are and where your competitors may be better than you.

Similarly, former CSIRO Chair Simon McKeon said startups that don't have a weather-eye on the wider landscape 'scare the living daylights' out of him as an investor:

> They're aware of 'a' problem, they're on their way to creating 'a' solution, but they are so consumed that what is often a fast-moving world is a bit of an unknown to them and their 'problem' could be solved before their product is in market. The last thing I ever want to do is take away the focus of the team that is cracking the atoms, they've got enough to do, but their wider team or even a board or advisors needs to be really connected to the world, otherwise it's going to have a risk element to it, which may be unacceptably high.[171]

It's a flaw that many scientists have: we forget that once we enter the market, we change it, and our competitors react to that change, so our market vision must remain fluid and adaptable as the market dynamics react to our presence. Semiconductor companies were invisible to Michelle's team and largely irrelevant because they weren't quantum – but we must never forget that the customer doesn't care what the technology is: they only care about the performance of the product. This can seem blindingly obvious, but we still manage to forget it over and over again as we fall in love with our own technologies.

There's a general rule of commercialisation that says whoever is closest to the customer gets to keep most of that customer's money. The further removed you are, the harder it is for you to capture value and the customer may not even see you. That relationship, that customer intimacy, is the next phase of what creates value in a startup after your unique technology; your customers really have to value you as unique. If you're not connected to them, if you're only selling as an original equipment manufacturer through others, then you won't develop that value.

A brighter future for scientist CEOs

Today when I give speeches or appear on panels at events, I focus on the role of technology and innovation in reinventing our industries and making life better for Australians and the world. I don't want to perpetuate the stereotype of that panel event early in my time at CSIRO (discussed in Chapter 1), where scientists are relegated to only speaking about what happens on lab benches or in our experiments. Not only do we need more scientists to become CEOs – we also need our CEOs to be louder and prouder about their scientific backgrounds, which I'm convinced are part of their success.

So, the next time you meet a mathematician, don't assume they're a professor — they might be the chief executive of Qantas, as Alan Joyce is. When you meet a computer scientist, don't assume the screen to which they are glued is full of numbers — they could be the Chief Executive of Netflix, as Reed Hastings is. Don't be surprised that the maths and science student became Chief Executive of eBay and Hewlett Packard Enterprise, as Meg Whitman did. Or that the Chief Executive of PepsiCo has a degree in maths and chemistry, as Indra Nooyi has. These are the skills that will steer our industries and businesses to success in a complex and challenging future.

Startmate

Niki Scevak founded Startmate on many of the same principles explored in this book, including the reason an MBA is not the right qualification for running a startup, the need for a clear market vision drawn from product management expertise, and the missionary passion that pushes through risk aversion, which Niki characterises as a naivety that hasn't been squashed by years of business experience. Under its new CEO Michael Batko, Startmate has gone on to become similar in some ways to ON and I-Corps (discussed further in Chapter 6), but in its early days, Niki was driven by purpose over the need for a formal program. He recalls:

> The original thesis of Startmate was helping nerds become great CEOs. It was the view that the technical founder or the product was the hero. If you had a great product, ultimately, you would win over time. The essential ingredient of a great company is a great product roadmap that can last many decades.
>
> There's plenty of stories where the product roadmap was great but the people were bad so the company implodes, but there's no case where you have this compounding over decades company without that magical product or the magical product roadmap. That was the hero starting point.
>
> So is it harder to get a product person to be a great CEO or to be great at business, or is it harder to get a great businessperson to be great at product? You can look at Steve Ballmer at Microsoft or John Sculley at Apple, or all of these other stories, where the businessperson learning product was not really that successful an outcome. It was always the belief that great product people would have an easier journey to becoming a great CEO. You know it's an unlikely story and you know that it won't work out most of the time, but when it does work out, that's when it's truly special.
>
> You observe Mike [Cannon-Brookes] and Scott [Farquhar] were just finished university and had no business experience; if you look at Facebook and Microsoft [founding CEOs], they didn't even finish university and now they're great CEOs; Google's [founding CEO] had just finished university. All of these super successful companies are actually staring you in the face that you should invest in these unqualified technical people and, if they do become great CEOs, then those are the best companies in the world.
>
> It's those unlikely, unqualified people, first time founders, first time CEOs, that's where the super successful outcomes come from. Not necessarily from a percentage of people that succeed, but from those who disproportionately account for the power law company. When you are successful, you have something to lose. When you have something to lose, there's something defensive that creeps in and something that holds someone back. It's like trauma or lack of naivety. We always discuss naivety as a positive attribute because it gets people to do unlikely things and run towards the opportunity. If you don't have naivety, you have, 'Oh my God, that's ugly, it's going to be hard, and I know exactly what's going to happen.' They're tepid or they're scared. If you have naivety, you just run straight into it. That's what's required. Obviously, it's like

kamikaze pilots, most of them die, but some win the war. That was very important. That was the central insight, we want to invest in product and technical people with no business experience to see what happens.

There is a chemical reaction where, if you believe in someone at the right time in their life, it just actually builds them. It creates something. That is the beauty of venture capital, when you believe in someone when they almost don't believe in themselves, or you nudge them to a higher ambition. That's the joy of the business when that reaction happens. Tyler Cowen is an economics professor at George Mason University in the US and said uplifting someone's ambition at the right time in their career is the greatest joy in the world.[172]

5

Market vision

When I returned to Australia in 2014, I realised that I didn't just have to challenge the idea that scientists couldn't be CEOs (discussed in the previous chapters), but I also had to challenge ideas about what science itself could and couldn't be. Australians were very comfortable with science pushing the boundaries of knowledge and making breakthroughs in 'blue sky' research driven by big-picture, open-ended questions. But at the time, we didn't, and still don't, think about harnessing science to its full worth to imagine the future – or going one step further and using those breakthroughs to actually *deliver* that future. Australia's scientific system is designed to encourage and reward a 'discovery first, application later' approach, and lacks what I call 'market vision'. In turn, efforts to bring the idea of a customer, or an application, into the research process have highlighted basic misunderstanding of how a market vision works.

Market vision can focus a broad field of 'possibly interested' customers down to a specific range of highly interested customers and use their insights to identify an opportunity for science to 'leapfrog' ahead of their current needs to drive their desire for a future state that they will help to co-create. This concept has driven the success of VC for decades and is not essential only for a would-be science-driven CEO, but also for any company seeking to stay ahead of disruption. It's how you cross the third Valley of Death (see Figure 3, page xii) and retain customers through disruption. Luckily for us, the breakthroughs and blue-sky research that Australia prioritises are essential features of market vision and actually position this country to race ahead of the competition by bringing the two together. But for Australia to turn its 'invention system' into an 'innovation system', we need to recognise and strengthen the role of science in creating a market vision and then turning it into a reality.

Market vision for deep-tech startups
Listen and leapfrog

I certainly hadn't come across any ideas about the 'market' when I was studying science at university. It wasn't until I went to the Valley that I really started to experience this concept. In the Valley, marketing is first and foremost about understanding how your product will fit into the market. What does the market need and what product is going to satisfy that need? And then you go even deeper, because at its heart, it's about predicting the future – something scientists do a lot of.

My experience in Australia has been that we tend to think of marketing as shaping our message and influencing consumers through communications, using words like 'positioning', 'brand', 'image', or even being used to describe just what is advertising. We don't tend to think of someone who does marketing as someone who actually has a vision of the market and can help you predict where it's going to go.

In Australian companies, using a 'market vision' lens isn't a natural skill in the way it is in the US and that's a big impediment for us. Tristen Langley said she learned her instinct for market vision from founder of major US venture firm DFJ, Tim Draper, who she said had 'done everything from seeing the future of the internet to seeing the future of each wave, each internet cycle and tech cycle, even crypto currencies. You need that big, huge vision and I think that's the challenge, translating that back to Australia … where the dream isn't always big enough.'[173]

So many business leaders and marketing professionals say, 'if you want to understand the market, just go and talk to your customers, they'll tell you everything you need to know'. While it's true that you absolutely need to *listen* to your customers, that's only one piece of strategic marketing required to create a true market vision. If your customers could tell you what the next breakthrough product would be, that product would have already been created. Learning from your customers is an absolute necessity, but it won't give you the market vision you're looking for. Listening to your customers is a critical step in creating the product they want, but if you only do that, which is what most companies do, then you get an incremental innovation. You'll develop today's product with a couple of extra features and at a lower price that you could have predicted without talking your customers. You have to use what customers tell you in a much deeper way to envision the future they want but can't articulate, because customers don't know what problem they're really trying to solve.

Spreets founder and CEO Dean McEvoy agrees that it's a poorly understood skill in Australia:

I once told a journalist there aren't enough product managers in this country, and they put a click baity headline up and people were like, 'Oh,

I'm a product manager!' I think the problem with product managers in Australia is they're really project managers. A lot of people who've worked in traditional IT or even managing software teams and stuff, think they're a product manager, but they're a project manager called a product manager. There's not this real science art form of how you elicit the insights that matter to someone enough that they would buy what you're building. That focus is not here that much, except for in successful startups.[174]

Bill Lanfri is a master at this. I met Bill when he was working for Accel and I was working at Iridex. We crossed paths when he was doing due diligence on a potential investment. Bill's background was in product management – creating and selling products for big companies, mostly telecommunications. Bill had the ability to really get where the customer's head was, really understand their problems at a deep level – to see it through their eyes. A really good product manager is like your customer – they've lived the customer's pain and they understand the things the customer is worried about. In addition to that ability, he had enough engineering expertise to know what solutions might be possible based on the company's R&D skills and could project forward to what it could invent that would delight a customer in ways the customer couldn't imagine yet. I had seen him come into many conversations about a product that wasn't cutting through with customers and help to completely change the direction of our thinking. He'd say, 'This is a mess, but let's take our head out of that and what if we had this? Something over here? Something out of left field that none of these customers had ever expected? What if we did that instead?'

MRD and PRD

Bill Lanfri, Peter Wagner (Accel) and Jim Goetz (Sequoia) all use a strategic marketing methodology to create a market requirements document (MRD), which spells out the dynamics of the target market: the market forces at work; the culture of the incumbents; the barriers to entry and exit; the sacred cows; who has religion and why; the uncomfortable truths; the gaps; and so on. An MRD is essentially a method to decide where the best problems to solve are and predict how the market will react to disruption. The MRD enables you to then drill down on a specific problem to solve and a product to solve it – that's when you create the product requirements document (PRD) to define that solution.

The PRD defines your startup and its mission – they are fairly easy to create when you are trying to displace an existing product with a brilliant science-enabled tech. The hardest PRD is for a market that doesn't really exist yet, as there are no comparable products to learn from. Unfortunately, this is most often where science and deep-tech needs to play. You can learn a lot from listening to market experts in AT&T, Intel, Caterpillar, Boeing, Telstra, Cochlear, RIO or BHP – large engineering companies can teach you how to do a PRD for a well-defined problem, but they call them different things. This is one of the reasons CSIRO works hard to keep relationships strong with these companies and create opportunities for small and medium businesses (SMEs) and early career researchers to access their expertise. Our partners at Sequoia wrote the MRD for next-gen wi-fi company Quantenna (discussed in previous chapters) and while the PRD changed as we learned more about the possibilities this technology would create, the MRD was visionary and so right that we never had to change it. Market visions from those same Sequoia partners created nine unicorns, three of which were US$10 billion companies – I guess you could call them deca-unicorns.

Large companies go wrong when they lock in the PRD and relentlessly try to deliver it by throwing the full weight of their people and budgets at getting it done – without iterating with their customers. Startups can disrupt them and win because they can be flexible, run 'what if' scenarios with customers, create rough prototypes to test and relentlessly iterate until they nail the extraordinary. Within those big companies you may find a Bill Lanfri, but chances are they have already left and are working with a great VC firm or personally investing.

Once Bill had that 'leapfrog' epiphany – one that would go above and beyond what customers said they wanted or needed but delivered in spades – he got customers to help him create that vision by critiquing it. They'd start to find the holes and he'd go, 'Okay, what should we invent to tackle that?' Before he knew it, the customers were actually doing the ideation for him to create the revolutionary product – something they could never have done on their own, but what he was able to do as a catalyst to drive it. In his own way, he was a great innovator, even though he never actually founded or ran a company for more than a year or so. That process of engaging with customers is what sparks off the neurons in your brain to get you thinking about how your technology can change the game for them in a completely different way. Good entrepreneurs really want to understand their customer's problem and empathise with them. Something in their brain latches onto a customer's experience and jumps forward to something that no one ever would have thought of.

When you involve your customers in ideation, you develop 'followship' – they see themselves as part of the development and journey of the product and your company. Followship is how you turn that small number of early adopters into a mass market that carries you across the third Valley of Death (see Figure 3, page xii). When we think about companies that have 'followship' among their customers, there are few brands as cult-like as Apple. Their customers love the company because the company loves their customers. Way, way back in the beginning, Steve Wozniak was a committed hobbyist who built computers in his spare time – he was the customer. Apple's original fan base included those people who loved computers enough to build them themselves. That's why Apple computers were designed to be visually stunning when you opened them up – all the wires were perfectly routed and the colours were beautiful. It was like a work of art inside and out because those customers used to love pulling the locks apart to see how it worked.

You won't have to convince the customers to ideate with you because they already want to do it. In the early days, Apple's customer following was cult-like because the customers felt like it was their company and their product and that they were part of the vision to take it forward. That customer-centricity was in the DNA of Apple and sent a signal to the whole company: we care about what we build. We're not going to ship anything that's not beautiful, that's not incredibly well engineered. You could see it reflected in the premium price their customers were willing to pay and the queues and hype around new product releases. It was because they were lovingly crafted.

Building up that connection with your customer pays off in so many ways beyond the original market ideation. Once you've grown and you're in the market, those customers will help you avoid your market share getting eaten up by a startup coming out of left field that you didn't hear about. While your customers can't tell you what to invent, the more you understand and care about them, the more likely it will be that your people will see and interpret the next leapfrog innovation idea.

I remember hearing from friends I had at Cisco about this crazy Australian company that had a weird new way of doing software. The Cisco engineers loved it, especially because it was priced so they could buy the licence on their company credit card without needing to get approval. The company was Atlassian, and its founders, Mike Cannon-Brookes and Scott Farquhar, knew the sweet spot to price their software because, much like Steve Wozniak, they had been those customers. They were two young software coders, with the universal coders' problem: collaboration through separation. They were simply looking for a creative way to solve the problems they were facing in trying to grow their team when most of their friends and colleagues had gone overseas, as so many Australians do.

They realised there was an opportunity in the emerging cloud technology to create a massive network of coders who could work together in real time without having them all in the same place if you had the right platform. It was a simple idea, but they had the perfect combination of being the customer and knowing their problem while leapfrogging the available technology to solve it in a unique way. They pulled it together just as big companies in the US were starting to develop cloud computing and make it accessible; and they had the insight to know exactly how to price it for their prospective customers and have it spread through word of mouth and sold with no sales force.

Successful deep-tech companies maximise interactions between their engineers and customers and promote peer-to-peer selling to embed that customer focus in their culture. Customers are not only the source of your revenue – they are also the wellspring of your ideas. As technologists, we often are fooled into thinking that if we simply create a better technology, the market will be ours. A business creates solutions for which customers pay. So, if better technology creates a better solution, then the world will beat a path to your door, right? Wrong! It's much harder than you think to displace an entrenched technology. You need substantial improvements, better cost structure, or both. Cash in the pocket is the customer's bottom line – if you grow theirs, they will put some in yours.

Startups succeed because they do something no one else had seen before, particularly in tech and science startups. Customers can never really have the epiphany to leapfrog products that already exist to develop a new innovation the way that the startup needs to, because they're so buried in their own day-to-day problems. This is why startups disrupt, because they come in with a wonderful solution that nobody ever expected and customers fall in love with it. They can fall in love so quickly that they abandon a 20-year relationship with a trusted provider like Nokia or AT&T and jump on board with Apple or Google or even a company they've never worked with before because the solution is so compelling. That is the magic of getting the market vision right.

Focus on your vision

When you see the world with a market vision lens, innovation takes on a more holistic, connected meaning: you see opportunities for new technologies to fill gaps in your vision; opportunities for old technologies to be reimagined in your vision; and you will need to abandon the old ideas that no longer fit your vision. Your market vision will be your guiding star when, as a scientist, you inevitably begin to see too many futures and need to focus in on just one.

New technologies

The greatest market vision person I've ever met was Steve Jobs. I was at CES (Consumer Electronics Show) in Las Vegas, literally the greatest tech show on Earth, when Steve came by the suite where we were showing off our laser TV. I'd met Steve before at a neighbourhood barbecue, and when he saw how our TV worked he got really excited – but not by the laser TV or the wireless tech. Instead he asked me what would happen if we shone the laser on a curved surface. Lasers, as coherent light, have the quirky property that they can be always in focus, meaning you still get a good image even projecting on a curved screen. Long story short, he wanted us to try to make a battery-powered laser projector for a small portable device he was working on. He couldn't tell us anything about this device, but his idea was to create a virtual keyboard and screen to significantly reduce how bulky a computer needed to be.

We tried hard to make it work but we simply couldn't meet the power budget. I drew the short straw and had to go tell Steve we failed – it was my last ever meeting with him in 2006. It wasn't pretty because Steve wasn't a fan of failure, but two things saved me. First, I had initially told him when we started that we could never do it – I had even mapped it out for him back in that suite in Vegas – but he still wanted us to try. Second, it turned out we had the same birthday – 24 February – and, coincidentally, that was the day of our meeting. The iPhone was launched the following year and I finally realised where that laser might have landed. The genius in the iPhone and Steve's market vision are self-evident, but so many of the enabling technologies like projective capacitance that enables multi-touch features came from small startups that Steve invested in and, in some cases, bought years before. That's the power of market vision – to see the potential in a technology well before it has reached maturity.

Better Place Australia CEO Evan Thornley and CTO Dr Alan Finkel saw well beyond the market vision of Better Place global's founder, software millionaire Shai Agassi. Better Place aimed to develop a network where electric car drivers could swap out their batteries as they ran low. Alan said the idea for Better Place came from a chance meeting between Shai and then-Israeli Prime Minister Shimon Peres at a conference to generate ideas for changing the world – but it wasn't a market vision directly connected with electric cars, per se:

> Shai Agassi got up and said the main thing that he envisaged is ridding the world of its dependence on oil. That was interesting. His vision wasn't to have an electric car company, or he didn't express it that way. His vision was to rid the world of its dependence on oil. As an Israeli, that would also be ridding Israel from the indirect influence that the oil-rich

Arab countries had that caused grief for Israel. Shimon Perez gave him a lot of support and introduced him to various people and that's what started the company.[175]

While the global parent company hit rough waters due to poor governance at its headquarters in Israel, Better Place Australia was setting out to shape its own destiny through Evan's vision for the Australian car manufacturing industry. Evan said:

> I was invited to join some of the top brass of the car industry at a series of major innovation workshops around 2010. Everyone had to put a Post-it Note up on the board of, 'My vision is, in 10 years' time, Australia's car industry will be …' People would say things like, 'Producing 400 000 units a year' and 'Successful and profitable'. No 'why' or 'how'. I proposed that Australia's car industry will be the world leader in large, powerful electric vehicles. I explained it was a valid vision for us because actually, we make large powerful vehicles, and the economics of large, powerful electric vehicles, particularly in fleet utilisation, were much more attractive than small city cars, which is why it was always weird to me that the rest of the world was going for small city cars. The industry adopted that vision coming out of that workshop. We got the support of Bosch, Continental, Holden and others to build the electric Holden Commodore as the proof of concept for how we would get there, which was a vision that was well beyond Better Place.
>
> I'm not pretending that operationally the entire industry was going to turn on a dime and do that. The much bigger problem was, of course, they were all foreign-owned companies, so the ability of the Australian executives to really influence the executable vision of the Australian industry was pretty limited. But there was real thought within significant leadership figures within the industry about Australia not trying to be the 14th country that builds small city car electrics, instead the first country that built large, powerful fleet cars.[176]

The vision was later brought to life by a number of other startups, including one called Rivian, which has been heavily invested in by Ford on the power of that compelling market vision, with a new partnership announced with Mercedes-Benz recently.[177]

Old technologies through new eyes

When you see the future through your market vision, you start to see new applications for old technologies. Sequoia VC co-invested with me on Quantenna at a time when semiconductors were completely out of favour. This was largely because of the cost to test and develop novel semiconductor technology, which was typically $100 million over 6 years to create a 'real' company. Meanwhile, the internet was thriving and you could create an internet startup for less than $1 million. The GFC made semiconductors pretty much impossible because capital dried right up and, worse still, the Quantenna idea was all about wi-fi, which had already been done to death, so what else could be innovated? We understood that innovation never sleeps and the 'what else' we saw was that smart antenna arrays and beamforming could be used to deliver carrier-grade transmission over a wi-fi channel that was considered fairly unreliable at the time.

The 'why do it' was much harder to justify. When there is an obvious market trend, like today's movement towards renewables in the energy market, the market vision is easy to create – but the competition is overwhelming because everyone else can see it, too. If you have a compelling market vision that isn't obvious and a market shift that others can't see, you incur more market risk. This is sometimes called a 'contra strategy', because it seems to go against the way the market is moving. But if you have a winning bet in a contra strategy, you will own the market and that model is irresistible to a good venture capitalist who understands the power of market vision.

The 'why' for Quantenna started with the average American family who would love to view high-definition video on four different channels on four TVs spread out across two levels of a large house. Say they call a provider like AT&T and order four set-top boxes, and about 6 weeks later a technician comes out and spends a day drilling holes, pulling cable and provisioning the home. But what if you could get one box delivered directly to the customer that they could just plug in and beam all the content anywhere in the house with no cable, no technician and no 6-week wait? It turned out that Quantenna's tech was so good it also worked in European homes with masonry walls and chronically poor wireless.

Similarly, Chair of Fortescue Metals Group Dr Andrew Forrest said he got into iron ore because 'it was an industry that had been so protected by established interests who never invested in anything new or built anything to grow the industry. No one believed it was possible to build the Fortescue Railway, but we said we would do it and we did.'[178] His career is full of breakthroughs in areas where people had given up looking.

Abandon old ideas

When you see the future through your market vision, you have to be willing to let go of ideas that no longer fit this vision and focus your energies on the ones that do. That's a lesson I learned when we merged my company, Light Solutions, with Iris Medical to form Iridex, discussed next.

Iridex

Milton Chang taught me to trust my market vision. He was the original investor in Iris Medical and the one who really backed the merger of my company, Light Solutions, with Iris to form Iridex. He could see the possibilities because his vision of the optics and laser industry was so specialised. He's the most successful investor I know – he's invested in 36 companies with no failures; all with his own money, not as part of a fund; and only ever in the optics and laser field. Milton had lots of reasons for his success, including his humility, the fact that he's always listening and rarely talking, and, of course, a deep understanding of how to analyse markets and people.

When we went into the negotiations to merge Iris and Light Solutions in 1995, everything about me and the Light Solutions team projected my belief in the uniqueness of our product and its role in transforming the optics market. On the other hand, everything about Iris said, 'We're already in the market, we've already got a product, why would we bother with you?' Iris were a great group and they really knew the market, but they were blinded by their own success. They couldn't see past the incremental innovation they'd made, or the trap of the third Valley of Death (see Figure 3, page xii) where their few passionate early adopters weren't becoming a mass market audience. Having Milton backing me helped a lot, but ultimately, I had to learn to trust my vision in the face of a stronger incumbent. It takes a lot of courage to stand your ground when you have a vision and say, 'No, you're not going to acquire me, we're going to merge and create a new company that's not married to any particular technology, either mine or yours. We're going to do what makes sense in the market.'

Iris was still a startup in its thinking, and it was disrupting the ophthalmic market with a new application for an existing infrared laser diode technology. Their product was building a good customer base, but it wasn't really unique enough to punch through. We thought that my green laser inside their product would be unique because no one else had my technology. The unique combination of my green laser in their product propelled sales 500 per cent higher and drove our strong IPO the following year. Two years after the IPO, we generated more revenue than the sum of all the prior years of each company.

After the IPO, the visionary former CEO of telecommunications company JDS Uniphase, Australian Kevin Kalkhoven, wanted to buy the company – he shared with us a vision of lasers transforming communications and liked our approach of doing with solid state lasers what semiconductors had done to electronics. Unfortunately, we could not convince the board to take Kevin's offer because they were so pleased with our IPO that they believed better things were ahead by staying our course. Kevin went on to take his company Uniphase to merge with JDS and become the most valuable company in the world – it would have looked like Facebook acquiring Instagram with similar financial returns for us founders. A few years later, Spectra Diode Labs (SDL) founder Dr Dave Welch tried to buy just the Light Solutions part of Iridex and again we could not persuade the board to sell. Not long after Dave's approach, Kevin bought SDL, making JDS Uniphase the most valuable company in the world. Either acquisition would have been transformational for Iridex and us founders. It's a reminder not to be blinkered by your vision because markets shift around you – boards and founders get comfortable after success like an IPO but you have to stay hungry and keep innovating. Even JDS and SDL were not immune from market shifts, as the tech wreck in March 2000 pretty much wiped both out given their customer base in internet companies. By contrast, as a deep-tech company that created unique value, Iridex continues today – this is the power of science to create sustained value, despite the volatility of capital markets.

Despite the good financial results of this merger, it was amazing how much resistance the original Iris team had towards the green laser innovation. Even when the green laser took off, people kept buying the infrared laser because it was cheaper, but the green laser tripled, quadrupled, then quintupled in sales and drove our share price. The infrared laser never really grew as much, but the company was obsessed with it because it was their original product, and it kept selling, just enough to stay alive. Even though they knew the ophthalmology market, it took me coming in from left field with this weird science and unique invention to disrupt their focus.

We carried both products for years; in fact the company still carries both, even though the green laser blasted through the market like a rocket. Subconsciously, I don't think the Iris team realised that they were always resentful of the shiny new thing that took off because they didn't invent it. Even a startup can be bad at disrupting itself once it gets going. At one point we developed the idea of putting a modem in the product so that it could 'phone home' when it needed servicing. But they argued that the infrared laser was so reliable that it wouldn't need to phone home. They weren't interested in the data this would give us, or how the customer would react to us caring enough to be tracking their product and giving them feedback. If I ever did a merger again, that would be the thing I'd look out for, because to be innovative, you have to always be willing to reinvent – to put the past in the past and move on.

I was frustrated that we weren't using the profits from the green laser to invest in something really unique. We managed to skunkworks something together using different contract-research funding and built a new application called Iriderm, a dermatology laser idea. We brought in one of my oldest friends, Dr Brad Renton, a fellow Australian and expert in the medical laser market. Iridem used a different type of laser technology to treat people with port-wine stains (birthmarks caused by capillary malformations) on their face, or telangiectasias (spider veins) on their hands or noses, safely reducing the pigmentation so you could hardly see it. We built a $10 million revenue stream company in the back room while the rest of the company was arguing about the infrared laser versus the green laser.

Support your peripheral vision

It's important to focus on your own market vision, but it's rare you will be able to achieve that powerful new vision on your own, so you'll need to support others who bring the missing pieces that your company can't provide. Sometimes you'll need to give something away to get something back; at other times you'll see opportunities to grow complementary opportunities.

Give something away

Uncle Frank Levinson always said you had to give away your best idea to make the vision a reality and he led by example. During the early years of Finisar, he developed technology that became the 'optical backplane' that's now a core part of every modern computer – a genius idea. But it was the classic example of an idea that came before its time: he couldn't raise any money for it. The optical backplane required a particular component that Frank had also developed, so he set out to persuade existing companies, with bigger budgets than he had, to invest in the optical backplane so he could sell them the component. While he gave away the idea that would have been a phenomenally successful startup, he did so to sell that component, which is now used in every backplane around the world. Eventually, he took that company public, achieving a multi-billion-dollar IPO.

In classic Uncle Frank fashion, when we were growing semiconductor startup Quantenna, we encouraged them to give away their best idea by building a miracle router that telecommunications companies could use, copy and roll out, giving Quantenna billions of sockets for their chips. It turned the few risk-taking early adopters into a mass market, crossing the third Valley of Death (see Figure 3, page xii). While chips sell at software margins, routers sell at wholesale hardware margins, meaning Quantenna made less bulk revenue but their profitability was extraordinary, so their PE (price to earnings ratio) was far better for an IPO.

In the early 2000s, I was recruited to be Executive Chair of a company called Intersymbol, discussed earlier in this book, which made a chip that enables you to encode information for telecommunications in a way that it would transmit further and faster but not be distorted. It added enormous value to a transponder, but unfortunately you had to give that idea for the transponder to the transponder company so that you could sell them the chip. They made 10 times more money than we did, but the transponder cost 10 times more to make, so on balance, while our little chip was the most profitable part of the transponder, they made more money than we did – but we wouldn't have had our success without theirs.

Complementary opportunities

Elon Musk knows market vision isn't a static image, but a living, breathing, dynamic ecosystem, full of cause and effect. It's rare that a market shift doesn't cause multiple changes. Really great entrepreneurs see this as a system-level shift and make multiple bets. If you think about it as a picture, then you do your roadmap assuming nothing else is going to change. But as soon as you start to operate, the moment you get in the market, you perturb it and it reacts to you; people will compete with you, people will go towards you and come away from you, other things get invented and change – your picture becomes a series of movies. So, your market vision has to be fluid right from the beginning. Really good strategic marketing people think about the counter moves the market is likely to make as they go through each stage of their evolution.

When I first met Elon at Stanford he was obsessed with solar power, electric vehicles (EVs) and space. Normally entrepreneurs are focused on one thing, but he strung the three together in his mind into one multi-layered market vision. Elon could see his vision and did an incredible amount of analysis to unpack how it would play out. Within the same decade, Elon founded SpaceX in 2002, became the largest shareholder in Tesla in 2004, and encouraged his cousins, Lyndon and Peter Rive, to found SolarCity in 2006, of which he became Chair. Each would go on to support the other in realising Elon's complex market vision.

SolarCity signed up residents to have solar panels installed for no initial cost, on the agreement they would buy back the energy they generated from SolarCity for the next 20 years. The company became the largest residential solar installer in the US, because Elon saw this fundamental market shift towards renewables coming, led by solar and batteries, and believed in it. He knew that once he created SolarCity, other companies through the supply chain would respond, shifting the market and dropping the price of solar panels and batteries. Two years after SolarCity was up and running, starting to create a drop in the prices of the batteries, a price drop that would also make EVs more popular, Elon became CEO of EV company Tesla. Soon, Tesla Roadster owners could access SolarCity charging stations for free and SolarCity became one of the first installers of the Tesla 'Powerwall', the battery installed in homes to charge owners' Roadsters. SolarCity made the initial loss needed to drive the drop in battery prices and, in 2016, Tesla repaid the favour by acquiring SolarCity in an all-stock US$2.6 billion transaction.

While SolarCity and Tesla are strengthening the appetite for batteries on Earth, Elon had his eye on a bigger prize: a battery-powered existence on Mars. The Martian day is 1 hour longer than an Earth day, but the power of the Sun on Mars is many magnitudes less due to the distance of the planet from the Sun and the frequency of dust storms on the red planet, among other factors. While SpaceX derives revenue from sending government and commercial packages into space today, its goal has always been to start a human colony on Mars living in its prototype 'Starship', which uses Tesla batteries. Tesla and SpaceX have also shared technological developments in areas like manufacturing, with both companies using 'stir welding' on car and rocket frames.

Elon looked at it as a whole interconnected system, stringing these things together in his mind to create a market vision. The more right you can get your market vision by making it fluid and responsive, the further you can leapfrog ahead of your competitors, because they're still seeing the static, two-dimensional picture, not the three-dimensional holographic video.

Power of paradigms

French novelist Victor Hugo is credited with the phrase, 'Nothing is more powerful than an idea whose time has come.' The corollary is also true and it's a trap waiting for entrepreneurs whose market visions aren't grounded in reality. A big, bold, ambitious market vision can deliver over and over again, but you must be able to see the steps to take from where we are today through to where that market vision comes to life. In particular, you have to understand the paradigms in which your customers are operating and what the barriers are to customers changing with the paradigm shifts you are trying to drive.

In the early 2000s, I co-founded and became CEO of a company called Lightbit, referred to throughout this book, commercialising technology out of Stanford. Our product was an optical chip that you grew on a wafer of lithium niobate – analogous to a semiconductor wafer – and could change the wavelength of a laser like a channel changer. I loved it because it was like applying semiconductors to optics. The inventors were excited about taking this tech to telecommunications companies because we were at the height of the telecom boom, and they could see there was a need in the market for a wavelength converter like this to operate inside the network.

The trouble came when we really dug deeper into the market. The need to change channels is clear and real, but the ability to execute that change inside the network was exceedingly hard because of the way it was currently built. You couldn't do it without changing the network architecture itself and that wasn't possible. This was a good lesson in realising that your innovation needs to be revolutionary, but is better if it fits into the existing paradigm.

What had looked so brilliant and so innovative at Stanford had been based on a detailed and thorough analysis of the market and the need for a wavelength converter – but it turned out to be all wrong because the inventors hadn't engaged the customer to think through how this would actually work and where it would fit into the system. While that seems obvious now, experts in telecommunications had invested and even joined our company, and it still took us 18 months in market to figure that out. But the upside of deep-tech is when one application doesn't work, you can pivot to another until you find success, as discussed in other examples through this book.

Dr Josh Makower co-founded a startup called Coravin, which commercialised an invention based on injection technology but applied it to taking wine out of a bottle through the cork, so it could be sold by the glass without wasting any wine through damage from aeration. At the same time we were investing in Coravin, the Australian wine industry was moving to screw caps – so why would I invest in something that depends on corks? Josh and I knew that, even though that paradigm shift was underway, expensive-wine drinkers still like corks, especially in the US, Europe and China. Most high-end restaurants today use Coravin, including in Australia – in fact, Australia became a pioneer in using the product just as it did for all of Josh's medical device startups. Using the fruits of this success, the company went on to create Coravin for Champagne and, of course, now for screw caps as well, because you win big if you can scientifically extract and maintain storage integrity. That win was all about understanding the business model of a restaurant and their customers – and whether or not they were likely to be part of a paradigm shift, or the exception that proves the rule.

At the birth of the internet, bricks and mortar companies were pronounced dead – but the future didn't play out that way because good companies adapted and played on the resistance to paradigm shift by digging in more deeply with customers who resisted ecommerce or whose businesses simply couldn't go online. Those who didn't adapt failed, but it took more than a decade to play out and that's plenty of time to adapt for those that wanted to. But at the same time, the ecommerce revolution wasn't always straightforward. Another of Josh's companies failed because it misinterpreted a paradigm shift. His company Nuelle was created to improve women's sexual health. The product worked exceptionally well but relied on distribution through digital marketing channels – online advertising through social media – to solve the challenge of women not wanting to discuss this issue with their doctors. Nuelle was intended for direct sale to consumers, bypassing awkward doctor conversations. As Josh explains though, Nuelle was ahead of the paradigm:

> One problem that I didn't realise, and the reason why the company struggled, is the major tech companies that control communication with consumers – Facebook and Google – view anything relating to women's sexual health as porn. We wound up getting blocked out of all the traditional marketing channels, regardless of our efforts to elevate the product as a health and wellness technology. We would say, 'Here's our clinical evidence, here's our highly qualified clinical advisory board, here's our completely non-offensive language, and our non-offensive imagery,' and yet we were completely censored. In the end, the company failed because we just couldn't get access to consumers. I really didn't expect that this was going to be our challenge. I thought, 'Forward-looking Silicon Valley, Google, Facebook will be all over this.' But they were just as bad, or worse, than traditional media in their bias against women's health.[179]

Josh came up against another paradigm shift with the sinus procedure company he chaired called Acclarent. The technology allowed ear, nose and throat (ENT) doctors at all skill levels to be successful treating sinuses usually only approachable by the most senior surgeons at academic medical centres. As these ENT physicians began to keep these patients, rather than referring them on, the academic surgeons saw the technology as a significant threat to their income stream. He recalls the challenges Acclarent had to overcome in overturning this paradigm:

> It really was an eye-opening experience for me, because I was shocked that physicians who I believed should be focused on caring for patients as a primary objective would try to find a way to destroy a technology that made the procedure safer and easier to perform. That really blew me away. These academics tried to force the company out of business by challenging the reimburseability of the procedure. By doing this, they hoped to prevent doctors from being paid for their services, and prevent the balloon procedure and technology from being covered by insurance. All this despite the fact that reimbursement codes existed that could be used and clearly were not specific to a technology.
>
> In the end, the doctors who had given the technology a chance saw for themselves the good results they were able to achieve in their own patients. These brave physicians combined forces and basically

threatened to leave the professional society and create their own. It was at this point that the academic leadership of the society realised they had gone too far. Eventually the leadership of the society conceded and the matter was put aside, at least for a while. Those community physicians saved the company and the technology, and preserved the option for patients far into the future.[180]

Knowing Josh, though, I suspect there was quite a bit of strategy in stimulating that bravery and awareness. Acclarent went on to be acquired by Johnson & Johnson and is part of its ENT division now. Dr Peter Farrell of ResMed said he had a similar experience when RedMed first started to sell its fully automatic sleep apnoea device: 'All the physicians were really upset, because they wanted to bring the patient back into the lab and set the pressure. Having it be automatic, that was going to cost them money.'[181]

While Josh was challenging a 'powerful physician' paradigm structure, on the technological side you could almost see Acclarent and one of his other companies, Willow, which made a wearable breast pump, as taking medical interventions back to an earlier paradigm – the one nature invented. The history of medicine is full of innovations and procedures that could be seen as technologies in search of solutions – the typical scientists' flawed approach. You can see this in the way many medical interventions actively work against the body rather than with it, including in invasive sinus surgery and bulky, awkward breast pumps.

With Acclarent, Josh's knowledge of the medical and surgical industries meant he could identify a need for a less invasive approach to sinus surgery, which for the past 30 years had involved opening up a patient's face and then stitching it back together. Acclarent always reminded me of the scene in *Total Recall* where Arnold Schwarzenegger's character removes a tracking device from his brain through a self-guided probe up the nose. That gives you a sense of how invasive sinus surgery is for this to be an improvement! I had to invest in that idea. Acclarent challenged the existing way of performing sinus surgery, which involved removing bone and tissue to clear the sinuses, by instead developing a balloon catheter, similar to one used in heart surgery to open heart valves, to just slightly expand sinus pathways and give the patient relief. Acclarent's solution could also be performed by ENT doctors and it didn't require surgery, so it reduced patient costs. That company went from idea to billion-dollar acquisition 10–15 years ago, when billion-dollar acquisitions were quite rare.

Similarly, Willow's less invasive approach meant it could fit inside a bra and won 'best in show' at CES in 2017.[182] Josh said he and co-founder John Chang developed the idea by starting from the point of wanting to do something for mums. They realised pretty quickly that the dominance of men in venture meant mums were an underserved market – further demonstrated when Josh said he didn't even know breast pumps were an issue until they sat a group of mums down to ask them about their needs and this was their number one issue:

> The inspiration for Willow actually comes from observing the micro details of the movement of the tongue and the soft and hard palates as a baby is suckling on a nipple. It was with that mechanism in mind that provided the insights needed to create the technology inside of Willow. Traditional breast pumps are huge, noisy devices, but in contrast, babies do the same thing, quietly and with their little mouths. We realised that there's some genius in this and we refocused on the details of how babies accomplish this so effortlessly and quietly. After studying the actual mechanics of how a baby feeds, we came up with a way to mimic those motions and pressures and were able to develop a technology small enough to fit within a woman's bra.[183]

Of his approach, Josh said:

> My favourite thing to do is to go after opportunities that everyone believes either are too hard, can't be solved or are already solved and thus they all look the other way. I especially like to chase after needs where everyone has given up, because while I'm working to solve those problems, I am likely to not have a lot of competition. The key inspiration for every single invention that I've ever been a part of is finding the small tweak that might allow the body to do what it is trying to naturally do anyway. In truth, I always take my lessons from physiology. I figure, after all this evolution we've got some pretty good systems in our bodies. When these systems get off track, it helps to understand how and why, and to try to consider what is the smallest input that can set it right. I'm always just trying to respect the body and understand how I can just give a little nudge to set it back. Normal anatomy and physiology are my greatest teachers and always show me the way.[184]

Tristen Langley is one of the early investors in a company that is still today a powerful example of the innovative David disrupting powerful Goliaths. In 2003, she was working in Silicon Valley and desperately trying to stay in touch with family back home in Australia, often using a ridiculous number of calling cards to make the international calls. A company that was then called Santa Cruz Networks pitched to Tristen when she was at VC firm DFJ with an idea to use VoIP (Voice over Internet Protocol) technology, offering a way to cut out the big telecommunications companies and connect over the internet. While there wouldn't have been enough Aussies in the US trying to phone home to keep this company afloat, this was also before the creation of the European Union, so you had no choice but to make an international call any time you wanted to call between the UK, France, Germany or anywhere in Europe. She single-handedly talked Tim Draper, founder of the VC firm DFJ, into investing personally – even though DFJ passed on it – and they were early investors in Santa Cruz Networks, a platform for VoIP technology. Diving further into the market, Tim and Tristen came across Skype (which was originally known as Skyper). Even though video calling technology didn't exist at that stage, the paradigm shift away from telephone calls to internet calls made the upgrade to video calls soon afterwards a simple step-change. Tristen said, 'The arrogance of the telcos was actually key. The secret of any entrepreneur is when bigger companies or incumbents are a little bit arrogant and cocky and they really think they've got the market to themselves. That's the best move for the entrepreneur because they can blindside the industry'[185] – a theme throughout this book.

Telstra Health

While deep-tech innovation needs more support in Australia, startups in internet or digital, as well as in healthcare, are growing strongly. However, former Managing Director of Telstra Health Professor Mary Foley has said innovation is still hindered by a widespread lack of understanding of how the healthcare customer paradigm works. For example, as a digital health startup, your customers are rarely actually patients – they are public and private providers of healthcare such as hospitals, government health authorities and clinicians in private practice. In understanding and navigating the complexities of the system, Mary said Telstra Health has grown from a business unit startup within Telstra to a subsidiary company wholly owned by the company. Market commentators value the business at over $1 billion – a unicorn inside an Australian corporate:[186]

> The way health systems work – and they have more in common in how they work than how they are different for about 80 per cent of what they do worldwide, though each also has its own idiosyncrasies – the consumer is usually not the payer for healthcare. The OECD average for direct payment by patients as a proportion of a member country's health expenditure is only 20 per cent. I see many academics (and others in technology) become very frustrated because they have good ideas, they develop new technologies and solutions, but then the people who are running the healthcare system tell them the health system does it already in some form, or that the solution cannot be integrated with clinical work flows or payment models or

with existing embedded technologies or that they are already
working on a commercial procurement process to market. That
nexus of being able to work closely between academe and
health delivery is often quite fraught and there aren't easy
mechanisms for making that connection.

People in the tech field (and many in medicine) become
very excited about the digitisation of the health sector, seeing it
as the next big sector of the economy to be disrupted and
revolutionised by digital transformation. For example, Eric
Topol's book *The Patient Will See You Now* describes how the
practice of medicine will be changed fundamentally by patient
empowerment and consumer-focused innovation arising from
the combination of digital technologies and genomics. In
healthcare, however, the transformation pathway is not a simple
matter of disruption and attempts to introduce digital health
solutions at scale often crash and burn.

All the American digital giants have tried to move into
healthcare in some form but their focus on consumer adoption
has generally fallen flat in attempting to bypass the established
systems. For example, some years ago one of the Big Five [US
tech companies] tried to create a personal health record that
everybody could use. Lots of healthy people thought, great
idea, I'll do that. But then, of course, the moment they're sick
and need to access the health system, that record is not
connected to or embedded with the health services that will
save their lives or make them better.

This doesn't mean that health systems can't change or
improve the way health services are delivered by adopting digital
technologies. But unless the technology innovator understands
all the dynamics of the health services ecosystem, disruption of
itself can impede rather than accelerate adoption. If you want
innovation to 'take' and to 'take' at scale, to really shift the
system, then you have to understand what problems you are
seeking to solve and where your solution is trying to fit in.

From my experience, it is essential to work closely with
health service providers (and their patients and clients) as key
partners. One needs to understand the core delivery systems
that providers require in their various settings – for example,
doctors' rooms, hospitals, pharmacies, multidisciplinary clinics,
aged care – and to consider what it takes to connect key aspects
of the patient journey across these settings (including with
clinical safety and data security). Working in partnership with
providers (and their patients and clients) and in the context of
the whole landscape can be the mechanism that can help to
bridge that Valley of Death.[187]

Telstra Health has created a role for itself as the connector across healthcare systems, academia and industry to facilitate innovation. Sometimes that innovation is through startups that Telstra Health acquires; at other times it leverages its significant platform to facilitate other partnerships. In this way, it echoes the contributions of 'cornerstone companies', discussed in Chapter 7. Similarly, its creation with investment from a corporate parent and research expertise from either academia or startup acquisitions is akin to the 'venture science' model, also discussed in Chapter 7. As outlined in media speculation at the time, Telstra Health faced its own 'near-death' moment around 2016 when Telstra reviewed the business and decided whether or not to keep it.[188] As Managing Director of Telstra Health from 2017 to 2022, Mary reflects:

> In digital health there is always that moment between the original innovation and the reality of implementation within complex health systems where confidence can falter. In my personal opinion, Telstra's decisions to continue to invest in Telstra Health beyond its initial acquisition and experimental phase shows that the parent company was able to take a longer-term view and support the mix of technology and health system expertise, working closely with the health sector, that was required to make the business successful. This included creating a separate company structure for the digital health business with its own board with external health and technology expertise.[189]

At Telstra's market briefing Investor Day on 16 November 2021, Telstra indicated that Telstra Health had become Australia's largest digital health company, including an expanding global footprint, with a trajectory expected to achieve annual revenues of the order of $500 million by 2025. Setting out to change the health paradigm from within is a bold ambition for a telecommunications company, but, as many others in this book have shown, often it takes a disruptive outsider with a clear market vision to be able to effect change.

Market vision at CSIRO

When I first came into CSIRO I had the makings of a market vision for the organisation – build a health team to address preventative medicine, build a digital team to help Australia catch up with where I had just come from in Silicon Valley, and create national missions to drive greater collaboration across Australia's innovation system to get a 'Team Australia' approach on solving national challenges. I also saw an opportunity for CSIRO to create a national accelerator and a VC fund (discussed in Chapter 6) so we could deliver more real solutions from science to create a new economic pillar to replace what we would lose from mining fossil fuels and to solve wicked problems like climate change. CSIRO had all the people needed to flesh out the detail behind these simple, high-level visions and together we made it real.

Scientists who hadn't come across the idea of a 'market vision' before initially found it hard to accept that, just like we predict the future every day through our weather or climate models, for example, we can use the aspects amenable to calculation to extrapolate beyond that to predict possible futures. It's a matter of creating enough different scenarios, using our understanding of the physical world and the interplay of the various elements with different models to get weak signals about the future that our intuition and other expertise can then latch on to.

The thing I love about this particular combination is that it's not all science – and exactly what our scientists were so deeply uncomfortable with. It's certainly not shooting from the hip – it's a combination of gut feel, experience and science that you put together to do a pretty good job of predicting what's going to happen. We have a real opportunity to not only reshape this thinking but also to take the idea of market vision to the next level of scientific rigour. Where market analysts often use quantitative measures like demographic statistics, interviews and consumer trends, we can add more depth to this by predicting the future using understanding of where science is going and how it will change the world around us. That field is yet to emerge, but AI will definitely speed it up.

So instead of trying to push the idea of market vision onto the organisation, I went back to the drawing board to think about how we could motivate our people to co-create and own this concept, and what our strategy team did next was ingenious. They reframed the market vision to be more like the UN's Sustainable Development Goals and asked CSIRO's people to work out what they would be for Australia. We said we couldn't do all 17 of them, so what would be the subset that really mattered? When we went at it that way, the scientists started to lean in, because it was quite a fascinating question. They did a beautiful piece of modelling to extrapolate what we knew then into the future. Themes started to emerge like food and water shortages, juxtaposed with an opportunity for Australia to produce and export unique, high-value foods that allow us to afford Australia's relatively high wages. We started to put all these pieces together and quickly came to the realisation that we had the opportunity to leapfrog in food. But we had to do things that weren't possible at the time and use science to unleash those possibilities in the same way you need to leapfrog in energy or in health.

Since coming back to CSIRO in 2015, I wanted us to be able to articulate in a simple and powerful way what its purpose was. At the same time as we were having this debate about market vision, we landed on our purpose statement: to solve the greatest challenges through innovative science and technology. This really resonated because that seemed to me why an organisation like CSIRO would be created. Then we asked, 'What are the greatest challenges we're trying to solve?' That dovetailed perfectly with the market vision/Sustainable Development Goals question. So, we distilled the themes from the market vision into what we called our six national challenges:

1. Our environment – how do we make sustainability profitable, so industry and environment become partners, not competitors?
2. Our food security and quality – how do we grow twice as much with half as much water and how do we shift from commodities to unique products?
3. Our health and wellbeing – how do we prevent more Australians from joining the 11 million currently suffering chronic disease?
4. Our future industries – how do we create new, high-value Australian industries and reinvent old ones so our children enjoy even better lives than we have?
5. Our energy – how do we navigate Australia's transition to zero emissions, without derailing our economy?
6. Our national security – how do we protect Australia from risks like cyber viruses and biological ones, so we can grow in both the digital and real world?

We intentionally kept the challenges fairly broad so that the full length and breadth of CSIRO people could see themselves as part of responding to these challenges, noting that the work being delivered under those broad titles is quite specific. Once we had these confirmed, we developed a series of industry roadmaps that have informed our partnerships and investments to keep us ahead of Australia's needs, as well as seminal reports like the *Australian National Outlook* in 2019 and the *Our Future World* megatrends report in 2022. That same year, the priorities outlined in the new federal Labor government's National Reconstruction Fund bore a close resemblance to CSIRO's six national challenges.[190] This alignment is great for the nation and shows we are starting to think more like entrepreneurs with a market vision and smarter about backing ourselves, and it's playing out in even more exciting detail in CSIRO's missions for Australia (discussed in Chapter 8).

Market vision in transition – mining

At the same time we were ideating around CSIRO and Australia's Sustainable Development Goals and we were co-creating a vision for the nation's future, there was a very real market vision collapse happening in one part of CSIRO – our mineral resources team. Towards the end of 2015, our minerals business unit was seeing its market contract by more than 50 per cent, led by the slow-down in Australia's resources sector, and it was desperate to find new ways to work with the industry to make it more environmentally sustainable and globally competitive. The team had been approaching things the same way for 20 years in the longest minerals commodities boom in history. You didn't have to do things too differently – just ride the wave – but it's a very different story when the market is going down. This example is a classic one for established companies.

They were the first CSIRO team to use market vision in a real way. The decline of the minerals industry in Australia presented us with a moment-in-time opportunity to ideate and co-develop many concepts with customers who otherwise would be too busy with production or too comfortable with success to spend time or resources on innovation. But it wasn't just good timing that drove these successes – the team had to completely change their usual way of interacting with customers. They'd been used to going out to the customer and asking what they wanted, or worse, going out to them and saying, 'This is what we've got and here's why you should buy it.' Instead, they transitioned to talking about what innovation the customer needed to be more productive and sustainable: 'Let's talk about how we're going to solve your problem. Let's do a series of workshops and whiteboarding and ideation, and let's learn together.' They talked about their ambitions for the next 10, 20 and 30 years.

You get an insight into market vision from customers, but it's not the whole market vision. When talking with Peter Coleman, then-CEO of Woodside, initially he was only able to tell us what he thought he wanted as a solution. However, when we took a step back and dug into what the problem actually was, that was when we were able to bring the teams together and really start to ideate. His vision was insightful: an oil rig built on the floor of the ocean that doesn't need all the infrastructure related to today's oil rigs, which sit on the surface of the ocean. You'd get rid of many of the risks of leakage, which is better for the environment, and have a far more efficient interaction – there's a whole bunch of reasons why that's a much better approach. It also seemed impossible, but that's another great reason to do it. While an ocean-floor rig vision didn't address all the challenges that the oil market is facing, it did recognise the importance of rare minerals in the ocean floor in supporting renewables.

You don't have to think very long about it to imagine all the other things you could do with a truly precise method of engaging with the minerals of the ocean, extracting them far more carefully and precisely in a way that would be environmentally far more friendly than it is now. It leverages unique communications technologies and unique materials. In fact, the results of that ideation session ended up in some of the work we're doing for NASA and the Australian Space Agency. It's a great thing to do, even if the ultimate idea may fail, because it can lead to all these wonderful nuggets along the way that can turn out to be very, very valuable, much like the first Moon mission led to many spinoff technologies.

It is these types of conversations that deliver far more value. This approach is also important because really novel science can take years to develop, so you want to pick really hard problems that are transformative to solve, otherwise it's not worth waiting a few years. It built a stronger understanding of why CSIRO was pushing to ideate 'extreme' science solutions, things that seemed impossible, rather than things that seemed fairly straightforward. Not only do those ideas take longer, but they can't be developed by CSIRO alone; they need to be co-developed with Australia's industries that bring the lived experiences and insights that turn a proof-of-concept into a seamless solution that can be operationalised.

With the minerals team facing their market shrinking by more than half, they decided to make a small number of very big bets. They basically assumed that everything they were doing now wasn't going to work anymore, so they picked a few big, bold things to work on and went really hard at them. Another idea to come out of ideation sessions with leaders of resources companies was what we came to call 'green gold'. We asked the heads of gold companies what their biggest challenge was, and consistently the answer was the use of cyanide and arsenic in manufacturing gold. Most Western countries won't tolerate the use of these chemicals anymore because of the environmental impacts; this limits their ability for onshore operations. Another question driving CSIRO is: what would it take for you to create more jobs in Australia? The solution in the gold industry seemed to be impossible: find a manufacturing process that didn't use those chemicals. An entrepreneur loves to hear the words, 'That's impossible.' Peter Thiele, the founder of PayPal, used to say you had to expose the 'really uncomfortable truth' that no one wanted to talk about and go for it. Elon Musk said the only things worth doing are the ones people think are impossible. If the idea is big enough, if you solve an impossible problem, the chances are it's going to create a lot of value.

Once we had our impossible question, we needed a partner. We started with one of the biggest companies in the industry. They could see how disruptive and powerful this idea would be if it succeeded – so they didn't want anything to do with it. If we messed with the gold process it would mean messing with their business model. The CSIRO team realised it made more sense not to approach the biggest company with the most to lose, but a smaller one with the most to gain as they would be far more willing to take the risk, with whom they successfully produced Australia's first gold using a non-toxic chemical process in 2018.[191]

It's a shame that many Australian companies so often think like the market leader we approached first. No company should feel like it's in so strong a position that it doesn't need to worry about disruption. The best companies try everything to not get that way, but once they get to a certain size, nature demands that they get disrupted so that the system can innovate and new ideas can break through. We can take comfort from the idea that disruption will come to any company when it becomes too big and powerful. Like in the battle between David and Goliath, the market will favour the innovative underdog that challenges the status quo.

The results of the mineral resources team harnessing market vision were extraordinary. They turned what had been a 20-year slow decline, followed by a precipitous market collapse, into the most significant growth they'd seen in 30 years. Their traditional revenue stream slowed to $25 million while their new model rushed up towards $50 million, which supported R&D that was directly benefiting Australian industry, improving its environmental and commercial sustainability. They also delivered their first ever IPO of a CSIRO spinout with gold analysis company Chrysos,[192] fulfilling precisely their market vision.

Market vision approaching transition – energy

Australia's energy sector is going through a massive shift. Globally, the world is shifting to renewables, both from a government policy perspective and in consumer behaviour. Market vision starts with asking, 'Where do we think the market is going to go?', which is usually a hard question, but in energy it's obvious. The world is slowly but surely turning towards a net zero future, but the details on how we'll get there are where there is opportunity, and indeed necessity, for innovation.

When we think about the energy transition, electricity represents only one-third of our problem and is by far the easiest area to decarbonise. The other two-thirds of our problem is related to fuels. Electricity and batteries are great at storing energy and giving you small amounts over a period of time, but they're terrible at shifting large amounts of energy quickly, delivering quickly or charging quickly. This is a real impediment for vehicles, for example, and for many industrial operations that rely on releasing massive amounts of energy quickly, a process that would blow up a battery or burn out a cable.

We looked at that challenge and asked what kind of fuel we could use that has no emissions and, ideally, is a liquid fuel able to use the existing car-refuelling infrastructure. We started to put our market vision together by imagining a world where you could fill up and run your car the same way you do now, but with no emissions. If we work backwards from that vision, how can science make that possible?

The characteristic of that process of market vision is always to ignore what science has delivered so far and just try to decide what would be the thing you'd want if you had a magic wand. Then work backwards from that to figure out how you'd get it. You start with the questions, 'What's the ideal thing that you want? What's my ideal market vision for the future where I know no one else can do this?'

This led us to hydrogen. It doesn't have any emissions, but in its pure state it's a gas and it's really hard to contain, so it meets only half of our criteria so far. Science has delivered hydrogen through electrolysis of water and fuel-cell cars, but how could we supercharge that? Water has two hydrogen atoms per molecule (H_2O), but ammonia is also a hydrogen-based liquid with three hydrogen atoms per molecule (NH_3), so what if we looked at that as a fuel instead?

The next step was to discover a way to switch seamlessly between hydrogen and ammonia, so you could have either one depending on what you needed. You could store it as ammonia, you could convert it to hydrogen, or vice versa. CSIRO found a way to do it mechanically with a special filter. This was a very different way of targeting science, because we started from the market and we went looking for the science to deliver the particular requirements that we thought would be the most compelling offer to the market.

Once we'd overcome the technical challenge, we needed a big resources company willing to support the infrastructure side and we needed to make it compelling enough for them to come on the development journey rather than be presented with a finished product straight away. CSIRO approached Dr Andrew Forrest and Fortescue Metals Group, knowing of his 'market vision' for renewable solutions for the resources sector, which is now formally the Fortescue Future Industries (FFI). Andrew explains:

> We've always wanted to be the example of a big industry that could go green without losing money so that others would have to follow our example. Fortescue has always been all about going green. At our heart is the pursuit of setting an example that we can do it profitably. That vision is paying off because, as the cost of energy has increased, we're seeing our competitive advantage increase from going green as it flows through to lower operating costs across the business.
>
> We'd been going after hydrogen from renewable energy as long as 10 or 11 years ago, but we didn't know how to store it and we didn't know how to transport it. It wasn't until CSIRO came up with the idea of the metal membrane – and now everyone's doing it, it's kind of old hat, but back then it was very novel. That gave us the plan B. It's not a particularly fabulously economic plan B, but it gave us the plan B that, if we can now store hydrogen as ammonia, we can transport it anywhere and strip it back if someone doesn't want ammonia, they want hydrogen. Then it's that plan B, which then triggered the entire Plan A of FFI.
>
> Science and tech has been my best friend here. There's not a snowflake's chance in hell that we could have hydrogen compete with fossil fuel. These are things which are unheard of. They come out of science and tech and I've had a lot of faith in

that, that the world will find the answers, we've just got to hurry it up. When you look at the savings you can make if the science succeeds, compared with the investment you make to give it a go, it's a no-brainer equation.[193]

Andrew remembers us discussing how to accelerate research in hydrogen, and I had asked him, 'How big do you think you can make this?' He said, 'A lot bigger than the Northwest Shelf', referring to the continental shelf off Western Australia that includes an extensive oil and gas region. I remember his reply vividly – it was the right thing to say. I knew his market vision was bright and strong enough to carry through what was promising but early stage technology to significant impact.

Together, CSIRO, Fortescue and others have realised Andrew's vision of a 'multiplier effect' that carries the innovation across many applications to start applying hydrogen to steel-making, remote mine sites the size of small towns, and ships to transition them towards renewables. No market vision is executed without learnings and failures along the way, and Andrew's has been no different, but his view on 'failure' has pushed through to subsequent success:

You have to have a crack and take a risk, and if it doesn't work, you learn, and either we will try again using those learnings or others will take them up and push forward. For example, we tried putting green ammonia into trains and found it didn't work as well as green hydrogen because it didn't burn as quickly, but that actually made it perfect for ships, because they burn fuel more slowly. Solutions uncover themselves when you keep trying new things.[194]

CSIRO has been uniquely placed to draw on its 100-year legacy of trust and delivering high-impact solutions to engage Australian businesses in innovative thinking around market vision. The companies that have been able to see numerous applications and payoffs are the ones who will secure their future and, I hope, will shift the needle for others through leading by example. Established businesses go through a 'J-curve' where they have to stop doing what they're doing and lose revenue as they begin to invest in something new, meaning they lose more revenue, but then ideally experience huge growth thanks to the new investment in innovation and become even bigger afterwards. But that's a scary proposition for a company that isn't currently in crisis or can't see disruption hurtling towards it. This is where scientists have the edge over CEOs from central casting, because not only can they see a brighter possible future, they can invent it – a scientist CEO can both invent it and deliver it.

Market vision for deep-tech venture

A market vision in the hands of an individual venture capitalist or a VC firm is a powerful tool to inform investment – and you need something to help you filter the 100 pitches you hear to distil the one golden pitch that you'll invest in. This section of the chapter looks at how a market vision can guide what to invest in, what not to invest in, and where geographically to invest to put your market vision to work.

Thesis-driven investment

Good market vision is what keeps top-tier VC firms ahead of the game. Really good funds string multiple market visions together, just as Elon Musk did in founding his interconnected and complementary companies, discussed earlier. When thesis-based investors go out to do clean energy investing, the good ones are going to be thinking, 'If everything goes solar, what are the other things that are going to have to change? Where are the bottlenecks? I'm going to go after and invest in them.'

'Thesis-based' VC funds form a particular market vision and go looking for investments that fit that vision, like Blackbird and its original digital market vision (discussed in Chapter 2). Blackbird combined the lower costs of building a digital or internet-based company with a focus on leading with the market to invest. In contrast, an 'opportunistic' fund invests in what seems like a compelling proposition, without necessarily forming a vision of the market. In the early days of the Valley, it was mostly opportunistic venture. The approach was to invest in the best entrepreneurs and the best deals while being ambivalent or agnostic to the market they were in. It was the job of the entrepreneur that you invested in to be across the market and 'figure that all out' for the company. While Blackbird originally set out to seize the digital opportunity identified by Startmate, Blackbird operating partner Robyn Denholm, who is also Chair of Tesla and the Australian Tech Council, said it takes a much broader approach to investment now:

> I love that Blackbird doesn't focus on one area or type of technology, as innovation can happen from anywhere. For Blackbird, with the community focus that we have and the connections in the community that we facilitate, I truly believe in the magic that does happen when different technologies/startups and people are exposed to different ways of thinking. From an economic point of view, I also like that we have many different technologies/business models/sectors that our digital native founders are focused on, rather than them being focused on thematics![195]

Over the past three decades, there have been more and more funds in the Valley and more and more specialisation. The most common fund these days is similar to Blackbird's original approach, focusing on internet-only, digital-only, or even one aspect of digital, like cybersecurity specialist startups. Australia's VC ecosystem is too nascent to have that degree of specialisation yet, and even Blackbird, which started out as digital-only, ended up becoming a bit more of a generalist fund. Even the highest degree of specialisation in Australia is quite loose and that's the funds that only invest in healthcare and medical, but all the other funds include health in their wide range of investments.

In deep-tech VC, I think you've got to have a market thesis to focus your investments and you've got to stick to that to achieve enduring success. Venture is about looking beyond the market that you can see today, beyond the obvious, and seeing the next thing coming before others see it. Market vision is even more important because of the short timeframes VC-backed companies are on. The first round of investment that a startup raises, an 'A-round', might last a year or 18 months, but you can't make a scientific breakthrough in that timeframe, so you have to be clear on where you're going and how you'll get there. If you're going to build your product with science and deep-tech, it might take you 5 or 10 years to do it, so you have to be visionary in your goal.

When we formed Main Sequence to manage the CSIRO Innovation Fund, we designed it to be specialist in two ways: first, it's a deep-tech fund, so it doesn't do internet plays or business model innovation, but instead focuses on translating research into impact; and, second, it shares CSIRO's national benefit mandate, so there is strong alignment with CSIRO's six national challenges, which I discussed earlier in this chapter. Both Main Sequence and CSIRO have access to a secret weapon for market vision ideation: the unique skillset of CSIRO's Futures and Insights teams, who combine economics with insights from scientific innovation to forecast where Australia's market will or could go. For example, Main Sequence backed a hydrogen startup called Endua (discussed in Chapter 7), which commercialised CSIRO technology. CSIRO ramped up its investment in hydrogen technologies years ago because we identified it in our energy roadmaps as a critical technology to drive the transition towards net zero emissions. If we hadn't invested in that early research, the company couldn't have been created for Main Sequence to back. The market vision really is a big differentiator for the venture fund because most funds don't have the depth of analysis of a national science agency that CSIRO can draw on to develop deep market expertise. In the same way that CSIRO has been more successful by running that way, delivering its first sustained growth in 25 years, I think the fund will be more successful than perhaps people expected.

The challenge for any fund in aligning around a market vision is when that can lead to total consensus across the fund's partners, because venture is all about making bold bets – and bold bets shouldn't create consensus: they should polarise people. Venture capitalists are betting that their vision of the future includes this startup; it might even be enabled by that startup. Don Valentine, the founder of US venture fund Sequoia, once analysed all that fund's deals to look at whether they were consensus decisions or split votes. He said that, by far, the deals that only one partner loved and everyone else hated were either their biggest wins or their worst loser investments. The trouble with consensus is you get to the middle of the road. If you're outside the box, you might get something great or you might get a disaster. The great venture capitalists have a gift for picking which one that's going to be.

Steve Blank said both the Lean Launchpad model and I-Corps have their basis in how startups build products.[196] Most of the leaders that made Accel a great VC firm, rivalling Sequoia and Kleiner Perkins, were product managers with profound market vision – Bill Lanfri (mentioned earlier), Jim Breyer, Jim Goetz, Peter Wagner and Joe Schoendorf all have this vision. Joe became a huge fan of Australia and comes every year for the Australian Davos Connection Forum. Rich Wong is now leading Accel, the first institutional investor in Atlassian. In the early 2010s, I shared a few quiet beers with Rich, Mike Cannon-Brookes, Scott Farquhar and Niki Scevak at a little pub on King Street in Sydney and it was fascinating to watch Rich navigate his way through the Aussie ecosystem. Niki and I were busy raising money for Blackbird, so it was ironic to be socialising with a US venture capitalist who had helped an Aussie company bypass the local VC system. Then-Managing Partner at Accel Jim Breyer was like many venture capitalists who didn't want their portfolio companies to be more than a 'bicycle ride away' (not that venture capitalists ride bikes) from the firm's office. But here was Accel's future CEO, exploring about as far away as he could get, drawn in by a great Aussie unicorn. It was a portent of things to come as Accel and the other top US firms overcame the tyranny of distance. We had done our job at Southern Cross in building a bridge across the Pacific, but it was time to build the next bridge at home.

Bill Lanfri had such a genius for seeing ahead of the market that he moved out of product management and into the world of startups. He ended up on a lot of boards, often being appointed by VC firms as a founding CEO to get a company up and running and make sure it locked in its market vision. You won't have heard of Bill, but you may have heard of the multi-billion-dollar public companies he's helped shape, like Bay Networks, SynOptics, Avanex and Jive. We really connected when we met so we got to know each other better over the years. In the early 2000s when I wanted to start my next company, Lightbit, he helped me engage Mayfield, which was the most prestigious VC firm at the time, and Accel, an up-and-coming firm. Accel's investment in Lightbit was their first in an Australian founder and got them interested in what else might be cooking in Australia that might be unable to raise funds at home; this is how Rich Wong came to invest in Atlassian many years later. Instead of making Bill our CEO at Lightbit, Accel put Bill on my board and he still mentors me to this day, as well as spending time with people from CSIRO, helping with commercialisation. It took me years of exposure to people like Bill who think differently to learn their ways and apply their lessons in my own companies as a founder and in investments as a venture capitalist. I liked Bill because I could tell he had an eye for market vision and I think he appreciated finding a scientist who was willing to try to understand it, because that's even more important than brilliant science or a brilliant invention.

Thesis-driven divestment

A market vision to guide investment is just as important to a venture capitalist as using the market vision to know when to get out. I've seen venture capitalists misread the market to the point where they think everyone getting into the market is a sign it's going from strength to strength, when it's actually about to burst. I've also seen them get so confident in a market that they thought they could drive the market with the sheer volume being invested and the influence that wielded over companies' direction. Leading up to the tech wreck, there was a period in the late 1990s where venture capitalists had convinced themselves that they had so much financial firepower that they could actually shift the market with their investments. If they went after solar, then solar was going to be big and everyone would follow them. That thinking drove a lot of the run-up in acquisition prices and IPO valuations, but it didn't last. Another example was in optical fibre communications. There were a dozen optical switch companies all making products for a huge market and every venture capitalist you talked to had done the analysis of how big the market was, but there was one fatal flaw: every one of those companies had only talked to the one customer. The venture capitalists thought that one customer was actually 20 or 30 customers. When they did their maths, they made the market much bigger than what it really was. Around a year before the tech wreck, they discovered they'd actually got this wrong, but then everyone was afraid to be first to jump. The market was riding high, but when the first one jumped, it was an avalanche that collapsed the market. Being blind to your competition or, worse, thinking that your investment is big enough that it's going to shift the market, or that you can out-spend the market, is a fundamental misunderstanding of the 'controllability' of the market. The market is going to do what the market is going to do.

I think Silicon Valley reinvents itself about once every 10 years, triggered by some kind of financial crisis about every 7 or 8 years as a rule of thumb. The first signs are when it starts to feel like you're at the biggest party in town and everyone is taking advantage of a golden opportunity. I've seen that pattern before when venture capitalists come rushing into the market – I saw it in the telecommunications revolution, I saw it in the clean tech revolution, and I saw it in the internet revolution. It's like starting off having a great gathering with a few friends, but before long, someone's put it up on Facebook and you get 1000 people and it's like a drunken party and you want to leave before you have to deal with the hangover and clean up the mess. It's time to get out and go to the next market.

So, at the same time as valuations are going crazy for clean tech or biotech or enterprise software, there is quietly something else going on here. The venture capitalists that are riding this wave of valuations are starting to quietly sell up and get out before the inevitable crash, and putting their money into the next thing. The really good ones like Sequoia or Kleiner Perkins have a gift for foreseeing that future market shift. I remember that just before the GFC, Doug Leone at Sequoia brought all of the Sequoia portfolio CEOs in and put up a slide deck. The first slide was an image of a tombstone and the next was a slab of raw meat. He told everyone that the steak lunches and good times were ending – he could feel it. He didn't know what the next wave would be, but he told them to raise their money, clear their debt, and get their businesses working properly straight away, because it was going to be a long time between drinks by the time the music stopped. It was extraordinary to see that slide deck leak out of Sequoia and move through the whole sector just as the GFC happened and the whole market went south. Having the insight to see the market ahead of time means you're in first to ride it and then to know when to get out. That's amazing genius!

When the GFC hit Southern Cross Ventures, we lost about half the portfolio in our second year. We did what all smart venture capitalists of new funds do: we invested in some later stage growth companies to convince our investors that they made a good bet with some quick wins. We threw in some high-risk (because they really seemed like impossible things to do) science-based investments as well. When the GFC hit them, as successful as they were with established revenue streams, the growth companies couldn't survive because they still needed a lot of capital to continue that growth. As soon as they were starved of capital, other better-funded competitors could just jump in and take over their market share. But the science-driven companies survived because they were unique and had no real competitors. That was a hard lesson for the fund, but a foundational one for me and my beliefs. Deep-tech investments aren't impacted by short-term economic cycles, but do have a cycle of their own that you have to watch, as discussed in Chapter 2.

Geography as market vision

After building Arasor in China, Silicon Valley and Australia in the early 2000s, as discussed in Chapter 2, I started thinking about the next market shift. If you could have been a VC fund in Silicon Valley in the 1980s or '90s you might be Sequoia today, but where was today's equivalent burgeoning investment hotspot? In the early 2000s, I thought Shanghai would be next and saw an opportunity for the Aussie founders we were helping to engage in the Valley to do the same in China and make a magic geographical triangle.

I had read about Chauncey Shey, the first investor in Jack Ma's Alibaba, the Chinese ecommerce company, and a mentor to Jack himself.[197] Chauncey was also friends with the founder of Softbank, Masayoshi Son, and had successfully founded Softbank China. I managed to get an audience with him through a complicated series of favours – including Dr James Zhang, head of China operations at premier VC Formation 8; my Arasor co-founder Simon Cao; Finisar co-founder Uncle Frank Levison; Dr Fred Leonberger, who was CTO of optics and telecommunications company JDS Uniphase Corporation at the peak of the tech bubble when it was one of the most valuable companies on the planet; and Vinod Khosla, a visionary venture capitalist from Kleiner Perkins and founder of Khosla Ventures. I was amazed by what a visionary entrepreneur-turned-venture-capitalist Chauncey was. It took three trips to China to convince him to partner with us at Southern Cross Ventures.

Around 2010, the Australian government sent several delegations to the Valley as part of their brainstorming to accelerate Australia's transition to renewables and I met with a few of them to talk through opportunities. Despite my inexperience with government and theirs with VC, after about a year of deep diligence the government formulated a new policy around innovation intervention in renewable energy. They talked with many people about ideas but ultimately settled on a VC fund, which they put out to tender for all of us to bid on. There were many groups bidding and the government planned as usual to hand out at least three funds and spread the risk. In the final pitch, I took a big risk and told the panel, 'Please don't give us this fund if you're going to split it into two òr three sub-scale funds. This will only work if it has critical mass of the full $200 million in one fund.' We had $100 million committed already but we needed their $100 million to match it. I stuck to my strategy and, although it would have been easy to settle for less, it wouldn't have worked, so why kid ourselves? The gamble paid off and in an unprecedented decision they awarded the whole amount to one fund: ours. Together with Chauncey and Softbank China, Southern Cross Venture Partners won the contract and established the Renewable Energy Venture Capital Fund (REVCF) in 2012.[198]

The REVCF was championed by successive ministers: first then-Minister for Resources and Energy Martin Ferguson and later then-Minister for Industry and Science Ian McFarlane. On the public service side, it was supported by Raphael Arndt, now CEO of Australia's Future Fund, Drew Clarke as then-Secretary of the Department of Resources, Energy and Tourism and Martin Hoffman, then the Deputy Secretary at the Department of Industry and Science, who would become the CEO of ninemsn and then the National Disability Insurance Agency. Little did we know at the time that years later when I became CSIRO's Chief Executive our paths would all cross again as Ian Macfarlane was CSIRO's then-Minister and Drew was then-Prime Minister Malcolm Turnbull's Chief of Staff, and both would go on to join CSIRO's board.

Chauncey did a lot on gut feel and, though he didn't know me at the time, he sensed a fellow entrepreneur. Over time we built extraordinary trust that enabled us to invest in some really cutting-edge Aussie science in renewable energy. We looked at everything from biofuels to algae to solar to laser-induced fusion. It was a major shift for Australia to expand innovation connections beyond Silicon Valley and into Shanghai and from there other parts of China. For example, startups XeroCoat and later Brisbane Materials Technology (Brismat) were technologies out of the University of Queensland to enable large-scale anti-reflection coatings to be easily sprayed onto glass – a great innovation for the building industry. We asked the team: what if we instead put it on a solar panel and increased the efficiency of a solar panel by about 3 per cent? It was a far cry from Professor Martin Green's groundbreaking solar cell discovery at UNSW[199] and the work Dr Zhengrong Shi had been doing there in the 1980s when we were students, but Australia had missed that boat and this was a great way to catch up. Brismat's technology could be applied to all solar types as well as beyond that into LEDs and complex electronics. Chauncey saw the opportunity to roll this out across China.

Another great Aussie deal was Ecoult, CSIRO's ultra-battery technology that was a self-healing lead acid battery for cars and had been acquired by US company East Penn when local Australian investors got into legal disputes with each other. John Wood, the founder and CEO of Ecoult, pulled off the heroic project of building 20 000 ultra-batteries to demonstrate grid stabilisation of the town of East Penn in Pennsylvania and proved the idea had legs. Sally Miksiewicz was East Penn's CEO and Vice Chairman, as well as the daughter of East Penn's founder and Chairman, DeLight Breidegam Jnr.[200] She was a visionary CEO and taught me a lot about the battery market, including her vision of using the CSIRO technology to stabilise renewables. Sally gave me the challenge of constructing a deal where an old-world family-owned business could co-invest alongside a foreign VC fund to solve a major problem for the world. Sapphire was the most ambitious deal we looked at, using technology from a series of US universities, national labs, and CSIRO's algae lab and library to understand the best way to use algae to excrete oil fuel. The common thread across all these investments was that, funnily enough, none of them were renewable energy deals to start with – like most university breakthroughs, they were solutions looking for a problem: brilliant science that needed help figuring out what application it could solve.

Building Southern Cross as a bridge from Australia to Silicon Valley led me to ask, 'What other innovation ecosystems should we build a bridge to?' When I imagined what it would have been like to tap into the formation of Silicon Valley in the '80s, I found a similar system in China in the 2000s. Partnering with Chauncey meant we had a bridge for Australian entrepreneurs not only to Silicon Valley, but also the fastest growing market in the world in China. We worked out of Chauncey's offices across China and he out of ours in San Francisco and Sydney. It was an extraordinary outcome, but the China–US–Australia connection was an incredible opportunity. That experience got me to rethink the role of government and turned out to be the first step down the path that would lead me back to CSIRO.

To be part of the foundation of the birth of a new ecosystem is another example of market vision, in that you should be there before it happens so you're ready when it does. The market vision of looking into China instead of Silicon Valley is an example of what an entrepreneur must always be doing – never get complacent with what is, but instead look always to what could be next. Today, I would suggest the market vision isn't China or the US, nor is it India – not to say you couldn't do well in any or all three of those markets, but I think those boats have already sailed. Instead, I'd be thinking about setting up my VC firm to feed the growth of new markets in Indonesia and Africa, and become the expert in them, the way you would have in Silicon Valley in the 1980s and '90s, China in the 2000s and 2010s and India more recently. Soon, if we pull a few more levers, I think Australia could have the next system ready to take off and draw more of the world's attention.

Australia's future market vision

Market vision is essential for a deep-tech startup to find its place in the world, and, as Archimedes said, from that firm place it can move the world. A market vision can also help a large, existing company to stay agile and responsive to its environment and can inform a policy direction that not only gives industry and research clear guidance on where to focus energies, but actually brings them together in a powerful, collaborative way. As a nation built on an egalitarian ethos and with safety nets that ensure no one is left behind, it seems counterintuitive to us to 'pick winners', but it's exactly what we have to do if we want our small population to take strides on the global stage. A clear market vision can set direction and inspire greatness, whether inside a startup or across a wide, brown land.

A scientist's guide for approaching a venture capitalist

Now you have a basic overview of how Australia's venture system has evolved, the value you bring as an aspiring scientist CEO (and your likely challenges), and a grasp of how to create your market vision, it's time to reach out to the most important person on your science startup journey: the venture capitalist who will not only financially support you, but also coach, mentor and guide you. These steps reflect how I worked with venture capitalists when I was a scientist CEO, and how I liked to work with scientists when I was a venture capitalist.

Do your homework

You will have studied the venture capitalist personally, their history and their current portfolio – you will know why you wanted to meet *them* specifically. You will value their expertise and experience. You will also understand that venture capitalists are constantly looking for the outliers that will make their firm successful. Each partner in a small VC firm typical of Australia (with a fund size of about $250 million) will consider 100 pitches and invest in just one, and they are likely to be managing 8–10 investments. Venture succeeds when one company out of 10 wins big and pays for the other nine, some of which die and some of which deliver only small returns. Logically venture capitalists should focus on that one in 10, but reality conspires to ensure they instead have to focus on the 'problem children' instead – this eats up an inordinate amount of time. So, you as a science startup will appreciate that the venture capitalist will not see you as the outlier that wins the day, and probably not even the one in 100 that they will invest in. So you will adjust your goals, as the next steps explain.

Don't pitch for their money – offer them something interesting

Build a relationship with a venture capitalist by inviting them into your lab every so often to have a look around, because good venture capitalists love to learn. They look at 100 deals and only invest in one not just because it means they invest in the best, but because they are naturally curious and hungry to learn new things. New things are where the breakthrough value is.

A lot of things conspire to distract venture capitalists from this meeting – knowing they've got to look at another 99 companies, managing the portfolio they already have and worrying about raising their own next fund. So make it easy for them to understand why you deserve their time by playing to your strengths and focusing on what's interesting – not what will make money. You are not the expert in what will make money. Focus on being yourself and find a venture capitalist who really wants to learn about your science.

First meeting is about asking for help, not money

Your first meeting isn't about pitching. You don't want money, you're not ready, you're not in a hurry – you just want to talk to them about it and see if they get excited. If they do – and they will because they want to learn – then you can start to pick their brains. You can start to ask them, 'What am I missing here? Help me. What do you think this would be good for?' You can shift what can be quite an adversarial relationship into more of a partnership.

Second meeting is about showing character and coachability, not CEO credentials

Science-based venture succeeds best if you can start on a collaborative footing. If you do it right, the venture capitalist will be interested, they'll give you some suggestions and they'll go away. This is the test: you won't hear from them again. They'll wait a week, 2 weeks and then they'll forget about you. But if, in that 2-week period, you reach out to them and say, 'I've been thinking about what you said and here's what I'm thinking, but I'd love to unpack it with you. Could we do a bit of whiteboarding on that?', they will love that.

They will see that you listen and you act. You're not just talking. They'll agree to come by and when you're on time for the meeting – because scientists have a reputation for not being on time – and you map out your ideas on the whiteboard and listen to their thoughts, they'll realise you're pretty smart and, even more importantly, they'll realise you don't think you know it all. They'll think, 'I can probably work with you because you might listen to me and I might be able to add value to you.'

The following meetings build the partnership – and you still haven't asked for money

You might have half a dozen meetings like that as the idea percolates and if they're really good meetings, you'll reject half a dozen business ideas as you crystallise your market vision: what's my product going to be; who's my customer going to be; and how is this going to make money?

You'll go through things and you'll realise things won't work and you'll go to something else – but you'll do this ideation. If it really works well, they'll bring in other people who they know can help you and you start to build momentum. That might take a year, but the beauty of all that is, you need it. Your company is unfundable without it. You now know an investor and you're kind of friends because you spent many meetings together over a year and they kind of trust you. It's a very different place to going in cold to pitch. I personally love that way of raising money, but you've got to be patient. If you're a scientist, then chances are you haven't got it all figured out, so that's a great way to do it.

OK, so when do you ask for money?

Let's start with things not to do: write a business plan, bring an 'advisor' or a banker, price the deal, use McKinsey or Porter charts, explain your startup's structure, give market estimates, say you have no competition, talk about paradigm shifts or platform technologies, or talk too much.

You'll be able to pitch them with a sentence that is so short it fits on the back of their business card if they give it to you (or yours if you have it prepared) or in an elevator if you catch them there.

When your punchy short pitch secures you the pitch meeting, you'll have fewer than 10 slides to run them through the six essentials:

1. Pain (problem) – who is the customer, what problem are you solving for them and why do they care?
2. Aspirin (product) – what's the solution?
3. Pie (opportunity) – how big is the market opportunity and how much can you really get?
4. Superheroes (team) – who are you and your team, and what qualifies you to take this market?

5. Superpower (technology or business model) – what's your unfair advantage or secret sauce and why doesn't anyone else have it?
6. Evil genius (competition) – who else is out there trying to take this market from you?

If I'm the venture capitalist, I want to hear all of these through the eyes of your customer. I want to feel like you are the customer, you've lived in their skin and seen through their eyes and felt their pain. When you talk about your competitors, do it through the customers' eyes. Don't fall into the trap of criticising them or pointing out holes in their tech – talk instead about what your product can do that theirs simply can't, and why customers will value the difference.

Now you've got their money – what next?

When a venture capitalist writes a cheque, they don't want you to just take the money and run. They want you to tell them what's going on and keep them engaged. Let them help you. Venture is very hands on, not like a traditional board (if you've experienced one of those). A good venture capitalist will spend a lot of time with you doing everything from strategy to sales. The more you let that venture capitalist into your business, the easier it will be for them to fight for you in their partnership for the next round of funding.

Who shares in your success?

Now this is hard to accept, but the value of your company has nothing to do with how brilliant your invention is, but instead comes from the economics of venture and how much money you will need to raise to deliver an exit – that is, go public on a stock market like the ASX. That's when venture capitalists make money on their investment in you, so that's what they're working towards from day one.

For early stage VC economics to work, the venture capitalist needs to own about 20 per cent of the company on exit. You'll likely go through three rounds of financing: series A to prove the technology and customer; series B to prove the product; and series C to prove the company. You'll typically need a new VC firm to price each round, adding to your first VC firm (not replacing them), so the venture capitalists end up owning 60 per cent at exit. Your team will own 20 per cent, and you and any co-founders will own another 20 per cent on top. For all the energy wasted on arguing, negotiating and trying to find the best valuation, it will still end up pretty much this way.

What you actually need to be focused on is how much money you will really need – I mean really, not 'conservative' projections – to get to the point of IPO. As a first-time founder, you probably have no idea how much you'll need, but a good venture capitalist will. All of this depends on you delivering the milestones you promised for each round, but also on the market responding the way you hoped – you might do everything right but still fail because the market shifted or a competitor came out of left field, or a pandemic or a war happened. Luck is just as important as skill in entrepreneurship – so you're all on an equal footing there!

Part 2: How to transform a company

The lessons of Part 1 don't apply just to scientists launching startups – they are equally applicable to Australia's largest and, indeed, oldest companies. CSIRO itself has been transformed by these lessons, from shrinking for 20 years to a new sustained period of growth delivering underlying positive EBITDA (Earnings Before Interest, Taxes, Depreciation and Amortisation – a measure of profitability) for the first time in its history. Corporate Australia can learn from the investment ethos of innovation and venture capital in tackling deep-tech and consider the traits that make scientist CEOs a force for change in the world – but, perhaps most powerfully, the corporate examples in CSIRO's market vision work deliver useful insights for a company of any size or age. We talk about Australia's three Valleys of Death as barriers to startups reaching sustainable revenue models, but there is a fourth Valley of Death that lurks in wait for established companies that become complacent – I think of it as 'commoditisation', and it's the dip that occurs during operational excellence and before the next disruption in the third Valley of Death (see Figure 3, page xii). Companies that don't continually innovate will become targets for competitors, especially hungry startups seeking to disrupt them. The fourth Valley of Death is when you get stuck in your leadership and stop innovating – that's why most job creation in the US over the past 40 years has come from startups, but that's also where the job destruction has come from, too. Part 2 explores the insights and lessons I've learned while heading up Australia's 100-year-old, 5500+ people-strong national science agency, CSIRO, during a period of significant policy experimentation, and draws on the experiences of our partners in corporate Australia through the model of venture science.

The programs CSIRO introduced as part of the National Innovation and Science Agenda (NISA) were designed to support scientists starting up companies, but they also set off cultural changes throughout the organisation that have made CSIRO more agile, responsive and adaptable. Conversely, the startups created in partnership with corporate Australia under the venture science model have not had to push change through large and complex organisations, nor deal with the cultural fallout of a merger or acquisition. Instead, through partnering with the research and venture communities, the corporates engaged in venture science have created new businesses that elevate their own market advantage and create revenue streams back to the parent companies, reminiscent of the formation of earlier Australian innovation bastions like ResMed and Cochlear.

The NISA initiatives were designed to bring scientists closer to the cliffs on one side of the Valley of Death, while the venture science model nudges industry closer to those on the other side. While making the leap a little easier for Australian scientists, the initiatives also incentivise companies through bottom-line benefits to their revenue and future financial sustainability, rather than short-term benefits like tax breaks or grants. Businesses that take a long-term view of their market and the opportunities created by science-driven innovation have always, throughout history, been the ones to outlast economic turbulence and shape the future.

6

Innovation agenda

My passion for solving Australia's innovation dilemma has steadily grown – from my first interactions with Australian entrepreneurs as an expat in Silicon Valley and then through my experiences setting up VC funds – to the point where I realised that to really make a transformative difference, I needed to be more embedded in the system. My time as a summer intern at CSIRO in the 1980s had shaped my understanding of the role of science to solve problems and it had always held a place in my heart. So I went after the Chief Executive role at CSIRO.

There were two key interventions I thought CSIRO could lead for the whole Australian research sector, addressing two of the major gaps in our innovation ecosystem explored through this book: an accelerator that supported more of our scientists to take their research out of the lab and see a future for themselves as CEOs of companies founded by their own technology, as discussed in Chapters 3 and 4, and a venture fund that would invest in the deep-tech science and CEOs to support them across the Valley of Death, creating that 'next wave' of venture discussed in Chapter 2. The government supported these bold initiatives, but it came at a time when innovation peaked – and then plummeted – on the national agenda. While that posed its own set of challenges, the fact that the nation has leaned back in to 'home grown' innovation again in the wake of a pandemic gives me great hope, because we have moved past national complacency into national interest.

Planting the seeds

If you look into the origins of many innovation systems around the world, you'll find significant injections of government investment helped kickstart their creation, like in the US where government agencies like DARPA, NASA and others have pioneered research in areas spanning space to clean tech. Government investment is at the heart of Mariana Mazzucato's *Entrepreneurial State* thesis, in which she charts the public investments in research that underpin everything from the various elements of smartphones through to the emerging tech heading up the NASDAQ today. Australia spends a lot on defence, although we will clearly never match the scale of US investment dollar for dollar, so it's probably not going to fuel an ecosystem on its own. But our national strengths in areas like agriculture, advanced materials from our history in mining, and in healthcare driven by our Medicare system, paired with our world-class research in these fields, are indications of where we might be able to compete, even with Silicon Valley. LookSmart and Better Place Australia CEO Evan Thornley suggests the role of government should not be as participant but as market designer – and to make his point in an Australian context, he explains it with a football analogy:

> What the AFL does is design the rules of the competition, so that a good game is played. Like having a salary cap. A well-designed market leads to a good competition. The trophies get spread around and that's good. In that case, the AFL commission has played the role of market designer. Government as market designer is a really helpful framework and government can design certain markets for certain things. Design a good market that is likely to lead to a particular set of competitive dynamics, and then get out of the way. That is a midpoint position, or an orthogonal position, to the 50-years-out-of-date debate of 'the government oughtta run everything' versus 'let the "free" market decide'. Every market has a design from accident or government decision, so let's make good ones.[201]

When I started at CSIRO as Chief Executive at the end of 2014, we needed a radical new strategy: what we would go on to launch as 'Strategy 2020: Australia's Innovation Catalyst'.[202] We launched CSIRO's first crowd-sourcing platform to drive a highly collaborative process, bringing in ideas and feedback from the more than 5500 people who worked at CSIRO. There were two elements of CSIRO's DNA that I wanted to make sure were front and centre:

1. *Customer first*: I wanted to elevate the customer voice to begin paving the way towards the organisation having a clear market vision (as discussed in the previous chapter) focused on solving Australia's greatest challenges.

2. *Innovation catalyst*: CSIRO mistook invention for innovation and so focused on licensing as the preferred path for commercialisation, and saw spin-outs and equity as a path of last resort. I wanted to reverse that and focus on higher-risk equity deals and create more companies – but not just from CSIRO's science. CSIRO often acts as a connector to bring partners together, so I wanted to amplify this role into being a catalyst for translation and commercialisation of great ideas and research into real-world solutions and products across the whole country.

As a result, Customer First, Collaborative Networks and Innovation Catalyst were among the key themes of our new strategy. It was both a response to shifts in focus I wanted to drive within CSIRO, as well as the significant national challenge I had come back to Australia and CSIRO to help solve. In recent years, VC funds like Blackbird, which I'd been part of creating, and incubators like Cicada had started to lay the groundwork for a growing innovation ecosystem in Australia. Building over the fallout from years of failed IIFs and waves of banker-driven VC, they chiefly led with investments in digital and medical fields, rather than in broader science-driven, deep-tech innovation and startups. I believed that CSIRO could be a cornerstone company (more in Chapter 7), an organisation to drive systemic change in Australia, just as Intel had been for Silicon Valley many decades earlier. Like a good catalyst, I wanted CSIRO to leverage what already existed and change the game a bit so that Australia could realise more benefit from its amazing scientific talents. I've always thought Australia needed something big enough at scale to be a catalyst. CSIRO was formed to deal with this fundamental problem of translating our amazing science into actual solutions to our challenges and creating jobs and economic growth, so I thought the next logical step would be for it to have an accelerator and a VC fund. While CSIRO is a government agency, I didn't think government should fund these programs the whole way through – they should be designed to figure out what industry needs, then go and solve those problems. They should be driven by market pull, not science push.

You also can't have innovation without failure. To innovate, you need a set of KPIs that are flexible enough that you can change and pivot and reinvent. That is a challenging proposition in a government agency, but we managed to make the case to stop playing at the edges of the problem and creating incremental programs – instead we needed to drive directly and deliberately into the Valley of Death. I think we won some early fans of our new programs by taking this bold approach – and also because the Valley of Death was not a place many organisations wanted to go.

National Innovation and Science Agenda

A few months after Malcolm Turnbull became prime minister, I attended a forum he was hosting with tech entrepreneurs like Seek's Paul Bassat and Freelancer.com's Matt Barrie at the University of Western Sydney.[203] I was asked to explain CSIRO's role as the bridge between research and industry, along with our plans for a venture fund and an accelerator. It was supported by Maile Carnegie, who was heading up Google for Australia and New Zealand at the time, as well as then-Chair of Innovation and Science Australia, Bill Ferris (discussed in Chapter 2 as the 'grandfather' of Australian venture). I was also busy working with Bill to draw on our experience in deep-tech venture to propose changes to the legislation around early stage venture capital limited partnerships (ESVCLPs) to make it easier for scientists to raise money, which would be important once the CSIRO Innovation Fund was up and running. The prime minister understood what we were trying to achieve and decided to fast track it so it would all happen immediately.

As the policy was being developed, we hosted a delegation in Silicon Valley, including people from the Department of Industry and the freshly appointed Assistant Minister for Innovation Wyatt Roy, who at 25 years old was the youngest person ever to be in federal ministry. I had a lot of fun introducing the delegation to people like Steve Blank, who invented the Lean Launchpad method and who started the US I-Corps program with Errol Arkilic, the architect of the original US Small Business Innovation Research (SBIR) program. I knew all these people from being in the Valley for so long and this delegation allowed the government to talk to entrepreneurs and innovators in another country.

Wyatt loved the idea of 'hackathons', sprint-like events where you bring together large and diverse groups of people to brainstorm solutions to difficult problems in fixed periods of time, like a concentrated weekend of activity. The term originated with events where computer programmers would 'hack' into a system to solve a problem, but it's broadened out of just computing now to include far more backgrounds and more generally means you 'hack' at the problem from different angles. So, we did a lot of them at CSIRO, pulling in the local entrepreneurs and venture capitalists and mixing them with the public service. If you're an entrepreneur, you might have a preconceived view of the public service. More Australians should see what goes on behind the firewall of government. Some extraordinary people have chosen to dedicate their lives to service – and it really is service. I loved watching my fellow entrepreneurs working side-by-side on a Saturday hackathon with my fellow public servants and only realising at the end of the day over a beer together that they were actually seeing the government from the inside. The work ethic, the smarts, the passion for what they do reminds me of missionary entrepreneurs. After all, innovation is about serving the customer, so entrepreneurs have a lot more in common with the public service than they realise and these hackathons proved it. We tackled my personal favourites: VC funding, stock options, litigation rules, insolvency, where the government should intervene, where it could cut red tape, and where it should get out of the way.

By December 2015, the National Innovation and Science Agenda (NISA) was a major economic announcement, made from CSIRO headquarters in Canberra.[204] NISA included 24 initiatives to support startups, reform company insolvency, and drive industry and research collaboration.[205] Under NISA, CSIRO would lead ON, Australia's first national science accelerator, and establish the government's first venture capital fund, which was also the first fund in Australia to specialise in deep-tech and science.

Of course, it was great to have these programs funded, but we knew it would take a bit of time before they were up and running, let alone starting to effect change in the innovation system. Where I saw an instant impact was in public service leadership. I'll never forget the first time I had lunch with the department secretaries at their annual retreat. I felt the atmosphere was very much, 'We'll be nice to this guy because he's new and he's running CSIRO, but, you know, it's just the same old CSIRO', and a few of them followed up after the retreat, but it took a bit of time to gain trust and credibility. Then NISA happened and people started to pay attention to innovation. It was extraordinary. Of course, it was still like any business relationship – you build up cred by delivering on what you say you will – but I've really enjoyed learning how the public service works and sharing some of my knowledge.

The ON program

Traditionally, scientists do innovation backwards: a solution without a specific problem it can solve. The ON program was designed to come at our commercialisation challenge in two ways – culturally, by teaching scientists entrepreneurial skills (reversing what they'd been taught before), and pragmatically, by 'teaching through doing' as they set out to figure out what their invention was good for and thereby commercialise their own research. It sped up the process of turning an idea on the lab bench into a product prototype and testing it with real-world customers to see if it had potential to become a company or pursue another commercialisation pathway. The intensive program taught scientists from across Australia's research sector about business model canvases, customer focus groups, engineering basics and marketing concepts, culminating in a 'Demo Day' pitch event for investors and other interested parties, teaching them how to commercialise their specific research, as well as any future endeavours. It ran for 4 years initially and was re-funded and expanded in 2022.

An ecosystem of deep-tech mentors

ON was officially created as part of NISA, but it was already operating in pilot mode within CSIRO, just for CSIRO researchers, during 2015; this made it easy to propose expanding nationally under NISA. We knew we had to bring in new people to operate it differently. But the danger of new people is they'll either polarise the system against them, or they'll try to too hard to fit in and won't make any changes, so it's a delicate balance. Interestingly, the architecture for ON was in part designed by a scientist named Sarah King, who I had mentored years earlier when she was working in a telco startup with Jonathan Lacey, who would go on to become an ON mentor. Sarah had found her way to a job at CSIRO before I arrived, so when I discovered her there, I asked for her insight to get us started. Topaz Conway, a fellow venture capitalist and investor who had founded Springboard Enterprises to support female entrepreneurs and businesswomen, introduced me to Liza Noonan, her Springboard CEO, and very graciously allowed us to poach her to become our inaugural head of ON. My gut feel was if you put a female entrepreneur like Liza and a female scientist like Sarah together, you'll end up with a very different picture of what leadership looks like (as discussed in Chapter 3). ON would go on to have many proud parents shaping it over its evolution, but Liza and Sarah were the ones who developed its DNA.

Liza and Sarah both knew this was going to be a disruptive program, so at first they quietly poked around, talked to people and found those who might be interested in joining in. Between us we then leveraged our networks to find the entrepreneurs who would be willing to partner with ON and form the mentor network that would be crucial to its success. Perhaps unsurprisingly, we found many people who had always been big supporters of CSIRO, who we'd built up trust with over many years, and who were fascinated to learn more about what was under the bonnet of the national science agency. It was surprising how many entrepreneurs, who we knew were passionate about Australian innovation, said 'yes' when we asked them if they wanted to be mentors in this program.

Being an ON mentor gave them the chance to see a lot of different research and potential investment opportunities that they wouldn't see otherwise, because it's very hard to see what scientists are working on at their lab benches, whereas this program curated this for them. They did it all for nothing, with classic missionary 'pay it forward' mentality, although perhaps with the knowledge that they might find something that they wanted to invest in, work in or even run – but it was more about doing something to make a difference. That group got up to about 300 people and they're still all engaged even today, 8 years later. After ON's initial 4-year run (government programs are usually funded on a 4-year basis), we weren't sure it would continue, and it wasn't until February 2022 that new funding was confirmed for it to return and expand. In the lead-up to 2022, I cannot tell you how often these mentors asked me when we were doing ON again or how frequently they told me there was a gap in the system without it, and I know many of them were actively advocating for its return. It was the best 'pay it forward' mindset I've seen in Australia since we created Blackbird and it makes me so proud to be an Aussie entrepreneur.

Once we had the mentor network in place, we had to get CSIRO comfortable with letting outsiders see behind the curtain. Initially, I think a few of our leaders saw ON as a chance to allow people in their teams with new 'crazy ideas' to get it out of their systems. What they didn't expect was how many of those 'crazy ideas' turned out to be successful business concepts and that those people weren't interested in coming back to their old teams once they'd had a taste of entrepreneurship. That was wonderful for the participants, but not so wonderful for the leaders who had inadvertently lost talented team members.

The most profound response to ON from within CSIRO came when we extended it into our environmental research, typically seen as work that doesn't get translated – often defined as 'public good' research. But what is missing from that simple definition is that the application of innovation methodology can help public good research deliver much greater impact than it would have otherwise. I remember receiving the following note after talking to a veteran climate scientist who had been with CSIRO for at least 20 years:

> When I went into this program, I thought I was going to prove it wrong. My objective was to show that it was all bull and it was a lot of rubbish and we shouldn't do it. And about halfway through, I got really scared because I felt like I was going to fail. I felt like, what am I missing? And then at the end of it, I realised, wow, this is a completely different way to think about my science and I wish I had learned this 20 years ago, because I could have achieved so much more.

I got a few of those letters and comments from people, making me realise that we had found a way to catalyse something unique in the soul of a scientist – the burning desire to make the world better with their work. ON helped shift some of the cultural behaviours that create Australia's first Valley of Death (the gap between high-NCI, low-TRL research and low-NCI, high-TRL research; see Figure 1, page x) and paired with a strong market vision, ON is also tackling the third Valley of Death (reaching customers; see Figure 3, page xii).

Teaching business to scientists – an exercise in reverse-engineering

Through their university education, scientists are trained to seek something new, aim to have it published in a scientific journal, and then move on to start their next discovery. While new knowledge is incredibly valuable, scientists are often creating solutions without problems. ON teaches them to take their solution and find its problem. Instead of moving on to discover the next new thing, ON encourages scientists to push their discovery further and look for a way it can make impact in the world. Rather than trying to rewire the training that's hard-wired into our scientists, ON pushes them to solder on another component to their wiring.

You can see the scale of the challenge ON takes on by looking at something as simple as the difference between a business proposal and a scientific journal paper. When scientists write a paper, it's a story. You start at the beginning, you go through the whole story of what they did and the amazing discovery and you don't find out what the results were until you get to the very end in their 'conclusion' section. But when you read a business plan, it starts at the end. It says what they're going to tell you, what they're trying to accomplish and how they're going to do it. They sell you on the solution before they get into the detail of how they'll do it.

Steve Blank said it could be painful – but ultimately an epiphany – when scientists were sent out to have customer conversations in the US I-Corps program:

> We taught scientists about the concept of 'customer development' by framing it as testing hypotheses – which turned out to be very useful for scientists. Scientists are likely the smartest person in the room, but they came to realise they can't be smarter than the collective intelligence of potential customers.
>
> What I realised after teaching this for a decade is that the smarter you are, the harder it is for you to understand the need to get out of the building. They've always been the person to come up with the answer, so it goes against their wiring to ask others.
>
> We taught them that commercialisation is not the same as ideation – that was the key insight we got across.
>
> I remember in our first class for the NIH [National Institute of Health] cohort, one of the scientists stood up to present the results of the 10 customer interviews we make them do before they come in. They said, 'Here's what we think and I'm just going to build this and it's going to be great.' I said, 'Did you talk to anybody? That that was the assignment.' They said, 'Oh, no, I don't have to, I'm a PhD, MD.' I said, 'Sit down, you don't deserve to present. Everyone else in the room is a PhD

MD, you're the one without any facts!' On the last day of the class, 10 weeks after, they said, 'I'm a PhD MD who realised I had no data at all on day one, and after 115 interviews, I've learned my first idea was the stupidest thing on the planet.'

Successful scientist-entrepreneurs are not ashamed to have those conversations instead of writing the 400th paper for their journal. They'll invest time in actually talking to people, in contrast to their peers who are just scientists who prioritise writing a paper or keynote for the next conference.

It's not that they're not published, they have great publication records. But they're investing time and resources in other places that you won't see these other folks in, because they're uncomfortable or don't know how to do it. I-Corps was the training wheels for at least teaching that mindset to the people it didn't come naturally to, was my conclusion.[206]

Fortescue Metals Group Chair Dr Andrew Forrest agrees with Steve Blank, noting that when he recruits scientists, he tries to break habits formed in academia of 'information hoarding' to score publication credits: 'It's "never be the smartest person in the room", and that goes straight against the grain when it's "publish or be damned" during your academic and your post-doctorate years.'[207]

Phil Morle was one of our mentors in ON and became one of our partners in Main Sequence. Phil had previously formed the incubator/accelerator Pollenizer in 2008 to support entrepreneurs, usually those who needed help to build a digital platform around their business idea. Phil said this was flipped with ON:

When I first started working with the CSIRO ON Accelerator in 2015, this was the start of this movement into deep tech, science-driven companies. This generally had the opposite problem to the world that Pollenizer began in. Very technical people needing to understand how to wrap a business around it. The first 'bootcamp' was a Tower of Babel with few of the teams understanding each other. Even scientist to scientist, they spoke their own language. But wow, when we understood what they were inventing we realised the incredible impact that was latent and needed to be unlocked. Five years later this system was transformed with a 'new habit' in research organisations that understood how to start a company. We now see a constant stream of companies rising out of Australian research.[208]

The individual university researchers loved ON, because it was the first time someone was coming to them saying, 'We get it. You've got this amazing thing, but you don't know what to do with it. That's okay! We'll work with you.' There was a lot of leaning in and very little friction with the professor level within universities. By the time ON completed its first 4-year run, around 37 incubators and accelerators had been created in universities and by other government departments. They usually only took teams from their own universities or departments, whereas ON drew from the full Australian research system, giving ON a strong pipeline. ON didn't care what university or state or government department team members came from; in fact, when a team failed, we didn't send them home – we combined them with other teams so when they came out the other side, they cared more about the product they'd been part of developing than the institution they'd come from.

Despite this classic Australian 'copy and compete' phenomenon, it was still nice to see many of them become part of the group of champions calling for ON to return. The metrics for their success were often, 'We produced 50 companies', but the company was a professor with an idea who filed a patent, paid $1000 and created a company. It was rare that they went anywhere because it's the wrong KPI (as discussed in Chapter 1). There's really no revenue model without a strong pipeline of successful startups, so they didn't last long. There are a small number of great incubators and accelerators in Australia that have stood the test of time, in addition to some exceptional ones that have sprung up with the second wave of VC, like Blue Chilli and Stone & Chalk. But the long-lived ones like Cicada Innovations in Sydney and BioCurate in Melbourne, which specialises in medtech, have done smart deals with their state governments to solve the business model flaw with subsidy. Cicada is my personal favourite because it treats all tech equally and has always loved science broadly. When it was first set up by founding shareholders the Australian National University, the University of New South Wales, the University of Sydney and the University of Technology, Sydney, it was intended to commercialise their outputs. When Petra Andren took over as CEO in 2016, she shifted the model to commercialise all Australian innovation, solving their narrow pipeline problem with a great collaborative approach. Sally-Ann Williams followed her as CEO in 2019, leaving Google after 13 years, and is working hard to scale their model with support from CSIRO and others.

While ON was designed to operate in reverse to the usual entrepreneurial trajectory – ON went looking for problems to solve rather than entrepreneurs who seek solutions to problems – the network of mentors ensured that entrepreneurial behaviours were learned along the way. Mentors often fell in love with different ideas or startups in the process of supporting teams and went on to become their advisors or board members, helping them secure funding or identify leadership talent. Once a scientist has spent 3 or 4 months working with an experienced entrepreneur, they quickly realise that person is better at being an entrepreneur or a CEO than they are, or, at a minimum, that they can teach them how to do that better. Some of the scientists wanted to be CEOs and we encouraged that, not just because it's important to experience it but because Australia actually needs them (as discussed in Chapter 3).

We also tackled the inherent fear of failure in science by putting teams whose business ideas failed into other teams to continue their learning and strengthen the other teams with diversity. Failure was expected and, in a way, celebrated. It was an important part of the culture. For example, the team working on a new graphene product discussed in Chapter 3, who were trying to find a real-world application for their unique substance, went through more than 10 completely different markets before they found a purpose for it that they could really believe in with drinking water filtration. It is times like these that mentoring relationships are built and the mentors figure out what role they're going to play, and the fact the team got through that and kept going proves their value.

Globally, the best incubator I've ever seen is ExploraMed in Silicon Valley. The Australian Startmate accelerator (discussed in Chapter 4) and CSIRO's ON program and collaboration hub/maker lab sites are sort of modelled on it, but interestingly have evolved naturally without explicitly copying ExploraMed. Dr Josh Makower founded ExploraMed and coupled it to the NEA VC firm, from which it receives $5 million to fund a wide funnel of ideas. When Josh believes he has something from ExploraMed, he supports them to raise their A round of funding, say around $20 million, and pays $5 million back to ExploraMed to support the next would-be companies. ExploraMed solves several problems. It enables multiple ideas to be prototyped and tested by leveraging all Josh's networks and the stable of now 10 mostly successful companies. It was his experience at ExploraMed that led him to share this knowledge with Stanford, creating a great pipeline of potential entrepreneurs from the Stanford Byers Center for Biodesign. Josh and Professor Paul Yock founded Biodesign in 2001, and wrote the book on biomedical innovation. There are similarities with CSIRO and ON and Main Sequence, but what ExploraMed doesn't have is a national science agency to lean on as its test labs, to solve hard research challenges, and to provide really deep-tech expertise – but it doesn't need it, because Silicon Valley has those elements across its ecosystem. This is why CSIRO is such a critical catalyst and the key accelerant to my personal mission of helping to solve Australia's innovation dilemma.

ON is similar to the US I-Corps program run by the NSF. Designed by Steve Blank, who developed the Lean Launchpad business methodology, I-Corps is recognised as one of the best science accelerators in the world. Steve said I-Corps came about when Errol Arkilic, who was overseeing the NSF's SBIR (Small Business Innovation Research) program, approached him to discuss using the Lean Launchpad method to give scientists better guidance on how to utilise their SBIR grants. Steve said:

> Errol told me that he thought we had invented the scientific method for entrepreneurship. Errol recognised that scientists were innovators, but that you needed to pair them with entrepreneurs to make a commercial company.
>
> By the way, this is what we get wrong – even in the US with our National Labs, we have all this great stuff and we think tech transfer is licensing the patents, without realising that's not how companies start. Companies start with an innovator **and** an entrepreneur.[209]

Steve said the challenge with I-Corps was that, while the first generations of the class had the full support of himself, Errol and others, by the time they thought about expanding it beyond the initial skills to start up their business to the skills that taught them how to scale, key people who were the innovators of the NSF I-Corp program had moved on:

> About 3-5 years after we started I-Corps, I thought, 'We got them to this point, but here's the 15 other things we need to teach them to scale,' and there was no one in place who was anything other than administrators and executors of I-Corps. I think the initial version of I-Corps got Principal Investigators maybe a foot or two into the Valley of Death, but it didn't teach them how to scale across it. I-Corps wasn't a failure, it got them to what I would consider in American baseball first base or second base, but there was so much more we could have taught them.[210]

Both ON and I-Corps have a strong focus on getting scientists out of their comfort zones (often their labs) and talking to real-world customers. By the time ON completed its initial 4-year run, our internal estimates within CSIRO calculated it had out-performed I-Corps, with ON graduate companies raising almost three times as much investment. While there are similarities between the two programs, ON has a unique advantage – ON had CSIRO's market vision, challenges and industry roadmaps (discussed in Chapter 5). ON also had the skunkworks access to CSIRO and its people, which can turn a lab bench idea into a prototype fast.

Success and the return of ON

I believe ON is unique in Australia's innovation system as well as in innovation systems internationally. When we expanded ON outside of CSIRO as part of

NISA, there were already a few tech incubators and accelerators in the market, and we had to think carefully about how we differentiated it from the rest of the system and not compete, but rather contribute. Some of them morphed a little bit to be more like ON and compete with us for funding, because most accelerators' business models involve taking a clip out of the company on the way through – except that startups don't usually have any money, so it's not a great business model.

After 4 years, ON's metrics were extraordinary; ON had delivered for both CSIRO and the wider innovation system. Between 2016 and 2020, the program trained over 1440 researchers from 39 Australian universities and research institutes, as well as 600 CSIRO people and nearly 200 additional participants from industry, government and the community. ON created 52 new companies, many of which ended up with people who weren't part of their original team, but who joined them through the program as their own teams failed. The diverse teams collaborating across institutions did better than those that came from silos of excellence. This is another secret of ON's success and why it needs to be a national program, not aligned with a single institution.

The real test for ON wasn't the culture change within CSIRO or the response from the research community – although these were powerful measures of success. Ultimately it was whether or not anyone would fund these startups and if they would grow over time, creating jobs. The answer has been a resounding 'yes'. ON graduates that have continued to grow and create jobs include telehealth platform Coviu, biosecurity solution RapidAIM, robotics company Emesent and livestock methane reduction solution FutureFeed, to name just a few. However, outside of ON's experienced deep-tech network of mentors, we still have more to do in engaging, educating and growing the venture community.

As the Australian regulations around employee share options are simplified and become more mainstream, I'd like to see CSIRO expand the ON program by building in more economic opportunities and giving participants the option to pool some of their equity, creating something similar to an ESOP for ON alumni, so they can give back to the community if they're successful. When I've talked to ON graduates, they have that same 'pay it forward' mentality that was so important in growing Silicon Valley, so it would be good to have a mechanism that allows them to do that as part of the program.

CSIRO Innovation Fund and Main Sequence

ON was designed to mine the gems of Australia's research sector, but to de-risk them for a risk-averse venture ecosystem, we needed a fund to bridge the gap for investors and tackle the second Valley of Death (see Figure 2, page xi). There was a lot of scepticism from stakeholders when we first proposed a venture fund. Australia had decades of IIFs that had mixed success; Bill Ferris and his IIF-supported successful investments, like Seek and LookSmart, were among the highlights. The IIFs were certainly focused on the right problem – the lack of risk capital willing to fund science across the Valley of Death – but Australia didn't have enough venture capital at the time to carry a startup through early growth, and so they inevitably had to go to Silicon Valley to raise money and then they never came back. My argument was a system level approach using ON with CSIRO's skunkworks to cobble together prototypes to feed the fund, and the fund to feed the traditional Aussie venture capitalists, which were generally only willing to fund more developed, less risky startups.

The venture community was divided over the idea of a government venture fund: some were supportive, whereas others wanted the fund run like an IIF, with the opportunity for them to bid to run it. That helped us work out who genuinely wanted to help and who was motivated by the wrong things. That was important because we were never going to successfully make the case that science venture will return a quick buck – it was always about intervening to fix a market failure. We explained to the venture community that our intent wasn't to muscle in on their turf, but to build a bridge by de-risking science startups to make them ready for investment by the broader VC ecosystem. We assured them that the fund would not have the money or the appetite to fund these startups all the way through. We committed that they would have every opportunity to co-invest with the fund on anything they liked – although, of course, that actually made them suspicious of investing in deals that were offered to them because they were used to Australia's competitive rather than collaborative approach. There was a bit of gnarliness in the market dealing with that reaction, but I'm so glad we built the fund ourselves because it meant we could operate as our own startup and hand-pick the people we wanted to run it. We could also create opportunities for CSIRO people to spend time with the fund and be exposed to another part of the system, stimulating different thinking in our people so that we could create more things to spin out.

I met the vice-chancellors of the Group of Eight universities early on to seek their support, engagement and endorsement of the CSIRO Innovation Fund. Professor Glyn Davis, then-Vice-Chancellor at the University of Melbourne and now Secretary of the Department of Prime Minister and Cabinet, was a standout supporter. He also kept his word that the University of Melbourne would invest in the CSIRO fund,[211] which meant a lot to me. I never expected the universities to invest in CSIRO's fund, but I did I want them to see it as their fund, not just CSIRO's fund – I hoped for a 'Team Australia' approach. It struck me that not only was the system hyper-competitive, but it also didn't have a very strong understanding of how funds fundamentally work. Just as we'd seen when we created ON and before that when I co-created Blackbird, the classic Australian reaction was to compete for a perceived finite pie of benefits, rather than to collaborate and grow the pie. It's funny because, for funds in particular, it really doesn't matter very much who manages it, because if it's successful, everyone who invested will make good money, so there was no reason to compete over it.

Balancing government and venture risk appetites

In order to operate in a more commercial way, we set up a separate entity to manage the CSIRO Innovation Fund independently of CSIRO and so Main Sequence was created. There are still many strong connections between CSIRO and Main Sequence, but from an investment decision-making and financial perspective, they are independent.

Picking the partners in a venture fund is like casting characters in a play. You need the banker who knows when to sell; someone with corporate experience who can keep it cohesive, like a managing partner; someone who has run a real business and can coach the teams; the entrepreneur; and the disruptor who can prevent consensus. The risk with a team like this is they're all different actors who think very differently, and some of whom might have no experience of doing venture this way, so they will have to relearn a different way. That kind of a team can really blow up. As a CEO, it's very hard to get comfortable with having five people who are that different working together on a team, and it's even harder not to interfere, particularly as a CEO who has run a fund. I don't think this could have worked for Main Sequence if it wasn't for a few key elements.

First, I know each of the partners we hand-picked very well, starting with Bill Bartee who I've worked off and on with for almost 20 years. If the CEO of CSIRO didn't have such long-term relationships, there would be an unmanageable tension. You'd have to fill the relationship with governance to make it work – and that would paralyse it. So, I go to some of the meetings, listen to the deals, give advice or suggestions – but despite my instincts, I try to never be the CEO in that room, because the danger of being the CEO of CSIRO is people just do what you say and the whole reason you want to do innovation is that you want to listen to what other people say. So that's been a tricky balance.

Second, Main Sequence and CSIRO have porous walls and full access, but it's give and take. CSIRO has a seat on the Main Sequence investment committee, which can veto a deal that is not in the national interest or could cause CSIRO reputational damage. The veto cannot be used simply for CSIRO's commercial advantage and Main Sequence doesn't have the right of first refusal on CSIRO's science. Practically speaking, we know there are few venture funds in Australia that are willing to invest as early as Main Sequence does, or which are focused entirely on deep tech, so cannibalising other funds' deal flow is not an issue. Likewise, Main Sequence is not committed to invest in CSIRO deals. We want them to invest in the best science, wherever it is. If we were a company with a venture fund, it would be much harder to say 'no' to deals coming out of CSIRO because a company would require benefit from the fund, which we don't. Part of the reason I wanted the blessing of the Group of Eight vice-chancellors was to help us replicate the way venture capitalists in Silicon Valley used to spend time engaging with students and understanding science when I was at Stanford, discussed in Chapter 2. Main Sequence has that same relationship with CSIRO and a growing number of universities, and it benefits all sides.

Finally, the relationship works because CSIRO and Main Sequence have alignment in their market vision and thesis-driven investment, as discussed in Chapters 2 and 5. Main Sequence has the same six 'challenges' in its market vision that CSIRO has, and Main Sequence leverages the market analytics, forecasting and science that CSIRO invests in those challenges. Most venture funds don't have the depth of analysis that CSIRO has between our economic forecasting teams and our industry-aligned business units, each with deep market expertise in its space. Many funds rely on their startup entrepreneurs to read the market, but I believe a fund has got to have a market thesis to focus its investments and stick to that to get genuine success. In the same way that CSIRO has been more successful than before by running that way, I think Main Sequence will be more successful than people expected. When we started it, we told our first investors in Main Sequence that it would make money, but at present it's a strong global performer – how's that for innovation!

A financial and national interest investment

Bill Bartee and I had gotten to know Hostplus, the hospitality superannuation fund, back when we were trying to get them to invest in Blackbird. At the time Blackbird was too small a fund for them, but we kept cultivating the relationship. Their Chief Investment Officer, Sam Sicilia, is a visionary when it comes to venture. While most of the traditional investor market was rubbishing venture during the GFC, Sam saw an opportunity for a contra-investment (similar to Raphael Arndt and Australia's Future Fund investing during the GFC, discussed in Chapter 2). Hostplus didn't invest in Blackbird's first fund, but it did invest in its second fund and now Hostplus is probably one of Blackbird's biggest Australian investors. Hostplus also has investments directly into deep-tech innovation, including two in fusion energy startups.

When we took Main Sequence to Sam, he liked the differentiation of our value proposition, but I think it was the national benefit that is in CSIRO's DNA that really connected for him. He takes his duty to his customers seriously – when you go out for breakfast with him, he talks and really listens to the waiters (and tips well) because they're Hostplus's customers. He doesn't approach his job like a typical transactional banker; he's like an entrepreneur whose goal in life is to serve their customers. So, when he took Main Sequence to his investment committee, he didn't only sell the financial angle – he also positioned the broader benefits for Australia of investing in venture:

> It helps, when you're an allocator of capital, that you have a young demographic. I can't shy away from that. Their runway is 20, 30, 40 years and therefore I could make an investment in fusion technology today without the need for it to come up good in 5 years. It might be 10.
>
> You must believe that it's going to give you a return, but what you can't do is separate that or divorce it from its other benefits to the current or future economy. You do it for the return because a company will be spawned and its value will increase, but that company will employ people, they will inject that money into [Hostplus customer industries like] pubs and clubs and across the economy.
>
> We realised that there's a whole suite of startup companies that started up before 2000 that failed and then tried again and again, but they were all going offshore – Atlassian went offshore, everybody went offshore – to get capital. By not being an allocator of capital at the right time, at the right place, you are simply saying to them, 'go somewhere else'.

> By investing in venture capital, we're contributing to building a venture capital ecosystem here in Australia and, by providing capital here to those companies, they stay here and grow here and build the industries that we're going to need for a prosperous future. If we don't do that, then we're telling them to go offshore for funding. Die, or go offshore for funding. Neither of those two scenarios - the company dies or the company leaves - helps Australia's future economy.
>
> You can't really think of it as saying I'm only doing this for the return, not for anything else. Or, I'm doing this for the impact, but not for the return. In fact, when you invest in venture capital, you get both a return and an impact, which can't be separated.[212]

We didn't really expect it, but we'd hoped he might use that argument and it's what started to get other investors interested. Perhaps it's because he's a physicist that he saw the world a bit differently. Sam studied to be an astronomer but decided to pivot to academia when he completed his PhD. An application for what he thought was a job teaching mathematics turned out to be a job teaching data analysis to business students, which he said was the first step on his journey into finance:

> The business students were totally disinterested in mathematics, other than to get their qualification if it was a mandatory subject and move on. As understandable as that might be, it forced me to teach math methods in a clear and concise and non-technical way. What I didn't know was that 10 years later, I would be consulting to superannuation funds using that same approach: take a complex concept and work out how to convince a trustee board of the merits of an investment without reverting to finance jargon.
>
> Andrew Goddard at Towers Perrin approached me for a role at his consulting firm, and I said to Andrew at the interview, 'I'm afraid you may have the wrong person, because I know nothing about finance. I'm a physicist by training.' He said, 'So am I - I'm a physicist from Oxford. I'll hold your hand for however long it takes you. All I need is people that can think.'
>
> One of my deputy CIOs at Hostplus is a theoretical physicist like me, another is a maths major. Does it help that you have a chance at reviewing some ideas and thinking about the possibilities? Clearly it does, because the scientific method is ingrained in us. Physics teaches you a discipline that you must test assumptions, or think about concepts for which there's a lack of data or that are remote, where you can't put a probe in to generate more data.[213]

It was ground-breaking to have Hostplus invest in Main Sequence because, while Australian superannuation funds have invested in venture for years, they've not invested in many explicitly higher-risk deep-tech venture funds.

They invest in venture with lower risk, reflecting Australians' expectations of their super funds. Perhaps Australians are more risk-averse about their super investments than in the US because our legislation makes superannuation mandatory, compared with the optional superannuation model in the US. As discussed in earlier chapters, where there are legislative requirements around an activity, like board membership, they can have a chilling effect on appetite for risk, so you have to look for ways to proactively manage that. Hostplus CEO David Elia was allaying latent fears that come with deep-tech investment when he announced their investment in Main Sequence, saying: 'This is not just important to us for the societal benefits that will be realised, but also the sustainable and long-term returns that will be generated and which are in the best financial interests of our members.'[214] Even UniSuper, the main pension fund for Australian universities, had never invested in venture until recently, whereas in the US its counterparts for Harvard and Stanford's alumni are the biggest contributors to venture, recognising the power of venture that comes from science. Vice Chancellor of University of Adelaide (and formerly of University of Queensland) Professor Peter Høj and I, as well as a number of others, have encouraged UniSuper to change this mindset. Just recently, UniSuper has invested in Uniseed,[215] the venture fund now owned by the University of Queensland, University of Sydney, University of New South Wales and CSIRO.

The Hostplus investment in Main Sequence was significant, but Australian investors really sat up and took notice when we landed an investment from Temasek, the sovereign wealth fund of Singapore. I believe Temasek's due diligence for the Main Sequence investment may have encouraged a few more Australian investors to sign up. When I was Managing Director at Southern Cross, we tried for a long time to get Temasek to invest in us without luck. Temasek's leadership found us fascinating because we were the only Australian fund operating out of Silicon Valley while going back to Australia to find deals, and they loved that differentiation. Bill Bartee and I spent 10 years working with them on various projects without them really investing in our fund.

It blew everyone away when Main Sequence raised over three times the amount of private capital as the government investment across its first two funds. This was particularly impressive as IIFs could only access government money under the condition that an equal amount of private investment was raised as well, and a lot of the time that didn't happen and funds failed to launch. Main Sequence started its first fund with $100 million – $30 million in royalties from CSIRO's wi-fi patent and $70 million in government investment under

NISA – and raised $140 million in private investment towards the end of 2018, creating a first fund worth $240 million. Building on this success, Main Sequence raised over $300 million in private investment for a second fund in 2021, bringing its funds under management to over half a billion dollars, from an initial government/CSIRO investment of $100 million. The Australian government committed a further $150 million to Main Sequence in February 2022, paving the way for its third and fourth funds, which were in development at the time of writing.[216]

Emeritus Professor Roy Green, Special Innovation Advisor at the University of Technology, Sydney, notes the scale of the challenge confronting Main Sequence when it was created, saying it 'played into an area that was underdeveloped and was very pioneering at the time. I think it lifted the game for everyone, especially given that it was connected with science and engineering.'[217] The combination of having access to the national science agency, including its people and facilities, as well as CSIRO's recent track record with spin-outs from ON and being active in the investor and entrepreneur communities, was a great advantage to Main Sequence. People thought something exciting and new was going on at CSIRO. We also took advantage of shifts in the market caused by record low interest rates, which meant investors had more money than they'd ever had before and not enough places to put it to generate decent returns. They were looking for something unique, different and possibly even meaningful, where they could get some scale in their investment, so they took a chance on us. Nearly all the fund's deals now have co-investors from other funds, including Australia's now-largest fund Blackbird, where we already had good relationships and trust. It came as a welcome surprise to also have top-tier international funds like Sequoia co-investing, given how hard we'd had to work to convince the local ecosystem. This collaborative co-investment in startups was a key metric for the fund because it was not created to be successful in its own right, but to lift the success of the system as a whole by encouraging more funds to invest in deep-tech startups.

In mid-2022, CSIRO looked at early stage VC investment (ESVCLP) across the Australian system to consider the role and impact of Main Sequence in the critical early stage support for deep-tech startups. Venture capital has grown significantly since 2015, but with only small shares of the early stage investment going into deep-tech areas like manufacturing-related (7.6 per cent), agriculture (1.2 per cent) and mining (0.5 per cent), particularly when compared to health (19.2 per cent) and information communication technologies, or what we'd call digital or software (40.2 per cent) during the financial year 2020/21. Zooming in on the Main Sequence contribution, their investments represented approximately 75 per cent of the value of early stage investments in 20/21 in manufacturing and agriculture opportunities. The story gets a little better when we zoom out to look at venture as a whole investing in deep-tech, instead of just early stage, but it demonstrates there's still a strong risk aversion to getting in when the risk is highest but most needed for deep-tech.[218] Between 2017 and 30 June 2022, the fund has helped to build 50 deep-technology companies, which have created over 1700 technology jobs. Every dollar invested into a startup by Main Sequence has attracted over $4.35 in co-investment from other venture funds, corporates, strategic investors and 'angel investors' (individuals who invest in startups).

Expanding Main Sequence

In the US market, it's quite common for the investors to drop US$100–$200 million into a startup once it kicks into its growth phase, giving the US market a real advantage as it supports these amazing companies and secures their presence in their economy. Temasek in Singapore do it, too. At Southern Cross Ventures, every time a company got to scale, we'd offer our Australian investors the chance to invest A$100 million directly and they'd laugh at us and say we were trying to sell them our bad companies. A couple of months before our portfolio company Quantenna went public in 2016, we tried to get our biggest investor to put $100 million in, but we couldn't get them across the line. If they had, their $100 million would have been $200 million in about 3 months.

If we can get Australian investors in VC to think more like their US counterparts, who look for opportunities to invest in companies directly as they grow, we won't have to go to the US for growth funding – which means more profits and more economic activity here. It's so important that we invest here and drive our own economy, rather than feeding Silicon Valley the way we have done for the past 30 years. I have reason to believe it's possible for Australian investors, because the Australian Future Fund does it. CEO Raphael Arndt said after a few years getting comfortable with US venture through a 'fund a fund' arrangement and investing through managers, the Australian Future Fund developed a co-investment program to invest alongside their managers like they were already doing in other asset classes such as infrastructure and property. They even managed to negotiate not paying fees on the co-investment, which Raphael said 'materially changes the pay-off profile'.[219] If a government fund can absorb that level of risk and negotiate that level of pay-off, it bodes well for other Australian funds to follow in their footsteps.

Strategically, I want us to continue to grow the number of big traditional Australian investors to anchor Main Sequence so that when we show them the next Quantenna or v2food (discussed in Chapter 7) and offer them the chance to put $200 million in to fund its growth phase leading into an IPO, they will compete with each other to do it. If we can pull that off, then we will have a self-sustaining engine running and we won't have to go to the US to grow our companies.

I believe one more big wave of Australian innovation successes could be all we need – but by the same token, one big push in the opposite direction could be enough to derail a lot of hard work. What keeps me up at night is that we will stuff up one deal. An investor may trust us and put the $100 million in, then it'll fall over, they'll never forgive us and that will jam the whole thing. I've raised a lot of money over the years, cumulatively hundreds of millions, but each time I get the same empty feeling in the pit of my stomach that I had when I got my very first cheque from an investor. I'll never forget them handing me that cheque, looking me in the eye with complete trust saying, 'I know you're going to make me money', while all I could think was, 'I hope I don't lose your money'. This is the dangerous part of the ecosystem because while everything's going up, there are no problems. But you're not dealing with a very mature innovation system. If investors see a bad sign, they can get cold feet very quickly and that can rapidly work against us. That's the risk we're taking trying to get the engine to run.

We had more Silicon Valley venture capitalists come through CSIRO than any time in our prior 100 years between 2015 and 2018, by which time Main Sequence was up and running and took over the role of hallway-strolling venture capitalist. There was a buzz in the place as the Valley discovered innovation in 'old school' industries like agriculture and food. CSIRO is a treasure in these areas and when you mix the oldest genetics group in Australia with the largest data science group – both housed within CSIRO – great things happen. The challenge was that it was almost impossible to get these deals invested in by Australian venture capitalists and I didn't want to lose them offshore. Similarly in energy, manufacturing, chemistry and environment, it was too difficult to get local venture capitalists. We needed to take the deals further up the TRL curve, but venture capitalists didn't want to pay for that investment, so while you're adding huge value to the startup by de-risking it, you couldn't recover the costs, let alone profit from it. This is part of the classic market failure of deep-tech – that's why CSIRO created Main Sequence, which is beginning to change these norms.

It's getting better, but still, none of the companies I founded in Silicon Valley over the past three decades would be fundable in Australia today, so we haven't come far enough. Why would they be unfundable? Because they were only 10 PowerPoint slides and an idea with no hardware to prove it. Sure, they were on a lab bench and in some cases I'd cobbled them together in my garage to prove they worked, but really they were an invention and a vision. Lightbit had nothing but a patent from Stanford and a vision, but it was in the telecom boom that led to the internet bubble. The companies I founded worked because the market vision was crystal clear – the venture capitalists trusted me to solve the tech, but they themselves could see the market vision. The companies also timed the waves of disruption well, so venture capitalists were hungry to jump on those big market shifts. We aren't there yet. CSIRO will keep helping move Australian inventions up the TRL curve and Main Sequence will keep bridging the gap, but I hope we won't be using our market vision to plug Australia's venture market failure for much longer. The opportunities are immense if we seize them, both financially for venture, and economically, environmentally and socially for the nation as these missionary companies grow.

After NISA

Many of the companies CSIRO took equity in after they went through ON, or invested in through Main Sequence, have delivered financial returns to CSIRO. Many also work with CSIRO's scientists to keep growing because there's a shared history and that's a virtuous circle of innovation. Six years later, we have an amazing revenue stream from equity and innovation that we've never had before. Across the wider organisation, this supports the continued investment in research that delivers impact and benefit to the nation.

In reflecting on his achievements as prime minister (2015–18), Malcolm Turnbull highlighted NISA as aligned with his 'values and priorities' as a 'radical, transformative economic policy'.[220] More broadly, he contextualised it in Australia-wide changes through to 2018:

> Whether it was the NISA or simply the fact that I spoke a lot about innovation, the 'ideas boom' I talked about did come to pass. In 2018, a record $1.25 billion was invested in startups, almost 10 times the amount invested in 2013. Venture capital investments reached $3.1 billion in 2018, double the amount in 2017 and up from $230 million

in 2013. This rapid growth in the innovation ecosystem enabled a massive growth in overall science investment when you combine the public and private sectors. In short, innovation became both a buzzword and a benchmark for governments and businesses – a theme continued throughout my government. We ensured more government data, especially geospatial data, was made publicly available than ever before and set up the Digital Transformation Agency to bring government services into the 21st century, as well as Data61 within the CSIRO, which has become one of the world's leading institutions for big data research.[221]

While there has been strong progress, there is still a market failure in commercialising science, as distinct from commercialising digital, as shown in the ESVCLP statistics discussed earlier. You only have to look at the small amount of capital that Mike Cannon-Brookes and Scott Farquhar needed to get Atlassian off the ground – Mike's dad loaned them $10 000 – before it was cash-flow positive. Right now, the Valley and Australian private equity and banking are all chasing digital; but Australian science is creating amazing breakthroughs across the board, including in digital. Digital science, on the other hand, is a different matter, and there are some great areas like AI and quantum that can feed the pipeline for digital commercialisation.

Former Cochlear CEO and former CSIRO Chair Catherine Livingstone was President of the Business Council of Australia when NISA was launched. She said telling a compelling, national innovation narrative needs to focus on actions that can be taken rather than trying to educate stakeholders on the whole system-wide complexity:

> If we've made a mistake as proponents of innovation policy, I think it's because we've made it too big of an issue and too complex. We've presented the whole problem: 'This is an innovation "ecosystem", these are all the elements', and we're talking to [stakeholders] who don't think in systems, and who are organised in silos, so they can't consume what we're telling them.
>
> We've made the problem to solve so big, that it can't be solved. So, in the background, we're trying to work with the systems view and work with the hotspots in the system… [but when we're talking about it to others], it's back to, 'the perfect answer is rarely the right answer.'[222]

Catherine's case that Australia doesn't have a strong grasp of how the parts of an innovation system work with each other was brought into stark relief for me a few years ago, when COVID-19 rapidly accelerated the take-up of digital tools and my family had our groceries delivered for the first time. The poor teenager had to get everything out of the car himself because we had to keep our distance and he had to stop me from taking it inside before he'd arranged it all because he had to take a photo as evidence it had all arrived. It struck me as funny because we'd been having online orders delivered for years in California and those companies had never added that step. If for some reason the package disappeared or went to the wrong address or got damaged by the rain, you'd just email the vendor and have it replaced. There was implicit trust, and the logic was that a customer would only get away with pretending the delivery hadn't arrived once, because if they tried it again, the vendor would be suspicious. It just seemed to work. I hadn't realised how significant the lack of trust between vendors and customers was in Australia, but I could see how it would impede the take-up of digital ecommerce.

When COVID-19 sent Australians into lockdown in March 2020, their search for digital solutions for their work and social lives accelerated both the social acceptance and widespread uptake of digital technologies by up to 10 years according to CSIRO research.[223] Some of these digital solutions have seen profound growth as a result of COVID-19, like Coviu, a CSIRO spin-out telehealth platform that saw 5-year growth projections materialise in the first fortnight of the pandemic. It went from 400 medical practices using its software to 12 000 in 3 months following the addition of telehealth to Medicare.[224] Its pivotal role as part of the government's response to COVID-19, both in terms of the health response and the economic narrative, has been part of a broader shift in the way innovation has been perceived as a result of the pandemic. The pandemic has been a shot in the arm to science, if you'll pardon the pun, as we've turned to our medical and research experts for solutions to the virus and to innovation more broadly to consider what kind of Australia we want to invest in rebuilding.

ON and Main Sequence received fresh federal funding in February 2022, and this time, with runs on the board and reputations established in the innovation system, I expect to see them bring together more people and perform better than over previous years because, as a country, we're more confident and we've learned so much about what works and what doesn't. We've done our experiments, trials and pilots and we're ready to commit more to improving our national innovation performance. Now we don't have to explain to the investor community, people can see successes in programs like ON and in Main Sequence, and I think the universities will really lean into this as an opportunity.

The creation of the ON program alongside Main Sequence – an accelerator connected to a fund connected to the massive R&D lab of a national science agency – has always been the recipe for success to me. It's similar to the way Niki Scevak created the Startmate accelerator first, followed by the Blackbird venture fund. As Niki has focused on growing Blackbird, Startmate has evolved under new CEO Michael Batko to support other parts of the innovation system in need. Niki applauds the work Michael has done with Startmate, noting:

> I'm almost embarrassed to say that I'm a founder of Startmate because the current CEO Michael Batko basically re-founded Startmate a few years ago. The team now is 18 people and it has its own identity. Startmate is an accelerator but they've also started fellowships, which are almost like an accelerator for people's careers, and helps them get a job at different kinds of startups at different points in their career. There's also a community of investors, supporting angel investors get started and get smarter. That combination of startups and investing in startups, career education, and then investor education, that as a trio warrants Startmate being an independent company.

Similarly, ON has spawned numerous partnerships and programs to support different parts of the innovation ecosystem at different times, and will no doubt continue to do so, responding to new needs as they are identified. I also hope many more ON graduates are successful enough to enter the investment pipeline for Main Sequence as it commences fundraising for its next funds. I'm proud to have been part of their creation and see the difference they've made.

However, I am still concerned about how we get the balance right in Australia between support for digital and internet-driven startups versus deep-tech startups. I'd like to see Australia really elevate and prioritise other parts of science that are intrinsically tied to the industries where we have a national advantage and global opportunity, like agriculture, manufacturing, space technology, telecommunications and robotics, for example. That's the next necessary push from my perspective – but again, so much of the foundational hard work has been done in recent years. The world needs more breakthroughs in these areas and we, as a country, need to figure out where we can focus and unleash our untapped potential.

When I started this journey moving CSIRO into the Valley of Death, I thought it would be a temporary change – that after maybe a decade, an ecosystem would form that would make it unnecessary for CSIRO to take on the role of innovation catalyst. But what I'm seeing now around the world, even in an innovation leader country like the US, is governments bringing in entrepreneurs to help them understand how to innovate. They are moving their national science agencies or national labs more deliberately into the market failure that is the Valley of Death, and I am starting to believe that the market failure is a natural consequence of the modern world. The energy transition to net zero is one of Australia's biggest opportunities to solve our innovation dilemma. It's a domestic market opportunity big enough to feed VC scale multiples and create some massive local companies – but I've no doubt there will be more. If the future of innovation that solves complex, global challenges that no single country or company can solve alone rests in the hands of powerhouse national science agencies driving collaboration, then Australia is in good hands.

7

Venture science

The long-term payoff from taking deep-tech startups across the Valley of Death is that we create large deep-tech companies here in Australia, growing new markets, employing Australians and delivering benefit to our economy, environment and society. If we choose to wait for Australia's deep-tech startups of today to be the behemoth Googles and Intels of tomorrow, we will be waiting a long time. We need solutions to embed deep-tech innovation in our large corporates today, so they can accelerate the development and delivery of solutions from science. In this chapter, I discuss what the creation of deep-tech 'cornerstone companies' can do for the wider innovation ecosystem. I also propose a new 'venture science' model of company creation, custom-designed for Australian industry, research and venture to co-create the next generation of global companies, made possible by the unique model of having a venture fund alongside the national science agency. The venture science model is designed to nudge our big corporates into growing the innovation ecosystems they operate in and being the 'training wheels' some of our deep-tech startups need, like iconic innovation companies ResMed and Cochlear needed before they became successful companies in their own right.

Cornerstone companies

In the late 1980s, I looked around me and was inspired by the virtuous circles I saw in Silicon Valley powered by the new value science created. Intel was the cornerstone of the Valley, inventing the silicon chip to create the computer industry. That invention planted the seed for what would grow into a thriving innovation ecosystem, where the returns on investment into deep science and technology research were reinvested in new inventions, and so new value was created. Those who made their first profits from Intel invested in the next generation of ideas, and so on until you could draw a line in Google Maps – created here in Sydney, coincidentally – to physically mark out the innovation ecosystem. In fact, the story goes that Intel nearly built a major presence in Australia before ultimately deciding on Israel. It knew Australia had awesome technical capabilities in materials and semiconductor electronics, which is one of the reasons we're so good at quantum technologies today; these may yet be one of our next big opportunities.

Those cornerstone companies became like big oaks that sprinkled acorns around themselves. If you look at the geography around where big companies like Google have grown, you'll find this huge distribution of startups that are like the offspring of the company. Scientists most commonly start a company after they've tried to convince their boss they should pursue a great business idea and the boss told them to get back in their box and just be a scientist. Scientists in the Valley don't like to be told that; it makes us want to do the opposite – and that's a wonderful thing for innovation. One of the best things for innovation is for those initially incredibly entrepreneurial companies, like Google, Cisco and Microsoft, to get really big and start to become like every other company. When companies start to get paralysed by bureaucracy, structure, policies and politics, people who loved the company for what it was at its core get frustrated, and it doesn't take much for them to get irritated and leave.

Those companies are a bit like what the NSW government calls 'anchors'.[225] Special Innovation Advisor at the University of Technology, Sydney Emeritus Professor Roy Green tells the story of Nokia's second life in Finland after the advent of the iPhone in 2007, enabled by its tradition of regular technology forecasting, similar to CSIRO's *Australian National Outlook*:[226]

> The Nordics have got into the habit of forecasting the future. They have to because they don't have any iron ore or coal mines that they can rely on, they've just got to rely on their own ingenuity, as do the Irish, Singaporeans, Israelis, and many other countries.
> I made a visit to Finland once when Nokia was imploding and Steve Jobs was running away with the prize. I was there in the middle of their forecasting or their foresight exercise, and I said to them all, 'Well, you must be very concerned – even a bit depressed – creating this fantastic, world-beating company and then seeing it get trashed.'
> They said, 'Nope, we're all happy because we are using the Nokia skills matrix and the ecosystem within Nokia to create dozens of new companies in health, biosciences, medtech and various other things. Nokia was our skills incubator and now we're ready for the next stage. Let Steve Jobs produce some phones, what happens when everyone's got a phone? Then he'll be in trouble.'[227]

I've worked on both the east and west coasts of the US, and my experience in Australia has shown it to be a lot like the US east coast in its approach to 'healthy leaving'. Employees leaving to do startups is part of the culture of the Valley, but it's not part of the culture of the east coast of the US nor here in Australia. Outside of the Valley, leaders generally don't like it if you leave their company – in fact, they'll often come after you if you leave. They believe they own everything that's in your head and they're going to prosecute you because you're competing with them, and they don't like that. In Australia the company owns what's in your head; in California the company doesn't own your brain. IP rights here lean more in favour of companies and employers than they do individuals who leave companies. Even if you can make a compelling case that the IP you use to start your new company was not created on your former employer's company time or with company insights, you need to be prepared to pay the legal fees that go with defending that and costs if you're found in the wrong.

When I left my job at Fibertek in Washington DC, my boss reacted the same way Australian bosses do, threatening me not to dare start my own company in the same field. But in the Valley, it's understood that people leave big companies and start small ones in similar fields; this is important for a thriving innovation system, otherwise you end up with big, stagnant companies running a monopoly. In Australia we deal with monopolies through various regulators, but in the Valley, the system is self-regulated so the best idea or the best company can thrive. You don't want your employees to all leave with your IP, but you do want to have 'healthy leaving', because that's how the system regenerates.

In the Valley, it's understood that big oaks grow tall and distribute acorns and that's just the virtuous circle of innovation. In Australia, if you were to leave a company in the medical device field, for example, and plan a startup in medical devices, you'd get sued. Your employer would theoretically own all the IP in your head – whether you invented it at work or in the garage on the weekend. You have to go very far away from the same market and even then, in countries like Australia and the UK and Europe, you can still get prosecuted. The Valley figured this out well and Australia needs to change to do the same. Obsession with IP is a flaw in multiple dimensions – it stems from academia where the idea is everything, even though an idea isn't innovation until it's a real-world solution. IP lawyers teach founders about the power of IP litigation so founders put that protection in their pitch, which is the last thing a venture capitalist wants to hear when they are trying to think big about opportunities for the startup.

Oaks foster the understanding that retaining talent in your ecosystem is essential to its long-term viability. When a startup went under in the Valley, which happened all the time, other successful companies – even if they were a fierce competitor – would usually hire all of those people and say, 'Let's make the innovation and market bigger together' and find a way to bring out their talents. When oaks negotiated with startups in the Valley, they understood they had to feed them and keep them healthy so the startups could give the oaks the unique pieces of technology that they needed and couldn't develop themselves. Oaks knew they needed suppliers to be successful to feed a competitive, innovative system. This is the total opposite of typical MBA practice, which tells you to play your suppliers off against each other to get the lowest possible price. When I became a venture partner, I had to try to sell a little acorn with unique technology to a big oak with the power to drive a paradigm shift that would enable it. Unfortunately, when we got into the negotiating room, I found that the MBAs had arrived in the Valley and none of the virtuous circle was true anymore. While it was good business strategy, it just wasn't good Valley strategy – or it hadn't been before. In the end, we reached a more equitable deal but it took me reaching out to the oak's founders to find fairness.

In Australia, we don't have that ecosystem – yet. Although, during the national spirit of collaboration unleashed by COVID-19, we saw flashes of insight into what that could look like for us. For example, sleep apnoea device company ResMed began repurposing some of its products to supply respirators to Australian hospitals as we faced an acute national supply shortage.[228] Similarly, Managing Director of Microsoft Australia and New Zealand Steven Worrall said supporting Australia's startups is something that companies like Microsoft 'have to think deeply about':

> For Microsoft to be truly viewed – or AWS [Amazon Web Services] or Google to be truly viewed – as working within the national interest, then there has to be examples of how we are investing in the local ecosystem. How we are helping to create the next wave of entrepreneurs, knowing that organisations that are less than five years of age are the primary job creators in the economy. This is a massive issue, especially right now, because as a nation, we're looking for ways to grow our way out of the pandemic. The digital revolution and the digital economy should be a big part of that and programs that help to get organisations off the ground should be front and centre. In fact, just this year [2022], we have created a digital native business unit within Microsoft ANZ. We're now collating and collecting a series of investments to build that business unit.[229]

I wholeheartedly believe CSIRO can be one of those big oaks dropping acorns – or eucalypts dropping gumnuts, if you like – that will germinate into that ecosystem through the partnerships it creates around the country. In November 2021, CSIRO and Google announced a 5-year, multimillion dollar partnership as part of Google's wider $1 billion Digital Future Initiative, investing in Australian infrastructure, research and partnerships.[230] This is an extract of what I said at the press conference:

> I first met Larry Page and Sergey Brin 30 years ago at Stanford. My first company went public the same year they founded Google, and our two companies are still next to each other in Mountain View California, although theirs is quite a bit bigger! What a profound impact science-led innovation has had on that small town, creating millionaires who go on to found more companies worth billions that employ thousands of people.
>
> Imagine if we could do that here in Sydney? Australia is at a pivotal point and facing profound challenges as a nation, seeking to build back better after the pandemic, and seize opportunities to lead on the world stage. Google's investment in Australian science will supercharge our emerging innovation ecosystem, which is passionate and world-class, but still small and fragmented, hindering our national delivery of real solutions from science.
>
> Innovation ecosystems form around cornerstone companies – in Silicon Valley and Israel it was Intel. I came back to Australia to run CSIRO because I think it's the cornerstone company to catalyse Australian innovation in science. As the national science agency, CSIRO solves the greatest challenges through innovative science and technology – but we don't do it on our own. No matter how brilliant the science, it takes a company to make it real...
>
> Both CSIRO and Google have significant strengths in AI, but it will take a network of partners across Australia to turn that expertise into real solutions, focused on the markets where Australia can win – from farming to manufacturing to environmental management. Neither CSIRO nor Google can do this alone. The world's best innovation systems have not sprung up overnight nor have they appeared spontaneously out of thin air. The cornerstone companies created the foundation, then their success planted seeds around them through investment and partnerships and networks that grew more and more innovative companies, until innovation became the driver of economic growth... The seeds we plant today will inspire our next generation of great minds to grow here. Our success will attract many more partners to join with us to help grow a better future for all Australians.[231]

As I've thought more and more about Australia's innovation dilemma over the years, I kept coming back to the question of a cornerstone company. Australia's largest companies are often more obsessed with not paying tax than they are with creating value – we're still enamoured with the 'get rich quick' phase of entrepreneurial evolution. My theory is if you spend your time worrying about not paying tax and preserving whatever you make, then you're going to be locked into a 'finite pie' mindset, trying to claw as much of it for yourself as you can. Whereas if you focus on growing the pie by creating value, financial rewards will take care of themselves. Individually we don't like paying tax, but as a nation, we accept that a safety net for our most vulnerable is important as part of our culture, so it shouldn't be as counterintuitive for us as it seems to be. Especially when you consider the long-term capital gain that when your company wins big, you only pay half the tax anyway, so we should see it as getting a great deal while also delivering a great benefit. Perhaps it comes from one of the industries that has driven most of our wealth for the past three decades. Mining is a very mature industry in Australia and those businesses know how to hold their positions in it.

But there are green shoots that give me hope. Dr Simon Poole is one of Australia's rare serial entrepreneurs, founding companies including Indx in 1995 to manufacture components for optical communications systems, which was acquired by JDS Uniphase for US\$6 million in 1997. In 2001, he founded Engana with Dr Steve Frisken to develop components for telecommunications systems, which was then sold to Optium in 2006.[232] In 2008, Optium merged with Finisar and it grew into about 600 engineers working in Sydney to develop the world's optical backplane (discussed in Chapter 5). Finisar co-founder Frank Levinson describes Optium as one of their best acquisitions – and suggests it's one of the reasons Australia should back itself rather than trying to copy Silicon Valley:

> I think the future will find that the different perspectives of different climates, geography, cultures, and such will make startups better to be in each place that can hold on to some of its uniqueness. For example, Australia should be a renewable energy powerhouse. How soon can the country go energy positive in a big way and then use that to profitably pull-down CO_2, maybe even break it down back into C and O_2 and such?
>
> One of Finisar's best acquisitions came from Australia, a company started by Simon Poole and their WaveShaper product; more than a dozen years later, I still encounter people who tell me how they used that product daily![233]

Finisar's presence and Simon's leadership supported autonomous vehicle component startup Baraja, funded by both Blackbird and Main Sequence, among others, and co-located with CSIRO in Sydney. Similarly, when 'Where 2 Technologies' was co-founded by Australians Noel Gordon and Stephen Ma, and their Danish collaborators Lars and Jens Rasmussen, in Sydney in 2003, its 2004 acquisition by Google to create Google Maps resulted in a few thousand tech jobs in the city. They are not quite tall eucalypts dropping gumnuts yet, but they show strong signs of life in our harsh environment, as eucalypts are naturally inclined to do. Building on the plant analogies of Dean McEvoy and Niki Scevak in Chapter 1, we can grow tall eucalypts if we just stop harvesting and exporting them as soon as they start to sprout.

Better Place Australia CEO Evan Thornley and CTO Dr Alan Finkel also managed to drop an acorn, without even being an oak themselves. Better Place globally was developing a network of locations where electric car drivers could swap out their batteries to keep their cars charged, when new technology came out of Japan for 'fast chargers', which supplied 50 kilowatts of power compared with the 2 kilowatts when charging at home. Alan and Evan believed these would be an important part of the electric car recharging mix, but global CEO Shai Agassi refused to engage with the new technology. Independently of Agassi, Alan and another Better Place engineer drew up their own specifications for a 50 kilowatt fast charger. The engineers didn't like the bulkiness of the Japanese and European fast chargers, which were so large because they used mains frequency transformers; instead the engineers designed their own to make them smaller and sleeker. They engaged what was then a small company called Tritium in Queensland that made components for cars used in the famous Darwin to Adelaide solar-powered car races. Under commission from Better Place Australia, Tritium made its first fast charger, although shortly afterwards Better Place Australia went into liquidation following the failure of the parent company. Alan said:

> We wished them luck and gave them the IP from the specifications, and they were fantastic. They got a bit of investment from the Queensland Government and others and turned it into a spectacular success. Tritium is now one of the world's largest – if not the largest – providers of fast chargers and has listed on the NASDAQ. That doesn't mean they've paid me or Evan any royalties, but we did it for love of country. I actually feel a real parenthood thing about it. I'm proud of what they achieved. So there was a spin-out result from Better Place Australia's vision, a vision that Better Place Israel headquarters did not share, in fact, actively disavowed.[234]

The Better Place Australia example shows you don't even need to be a large company to be a tall oak (or eucalypt) – it's all about believing in the power of deep-tech innovation to grow the pie for everyone.

If you position your vision right and you understand them well enough, you can go to a big US company even as a small startup and convince them to support cutting-edge technology that's still in the process of being invented. I don't think that's yet possible here in Australia in the way it is there. But to be fair to Australian corporates, Telstra and the Commonwealth Bank invested in quantum technology research and technology at the University of New South Wales in 2015[235] – what CEO in Australia ever invested in a product that's not going to happen for 20 or 30 years? When have we ever seen that before?

Australian corporates are changing, which is a good sign, but the quantum investment was also significantly de-risked by an initial, larger federal government investment as part of NISA, showing the ongoing initial need for intervention in this market failure. If those corporates follow on with more investment it will be a sign of real change, beyond what a venture capitalist can fund over the 10-year life of a typical VC fund.

Venture science

To create cornerstone companies in Australia's risk-averse corporate landscape, we have to find a new way for Australia's big companies to engage with deep-tech innovation. When CSIRO founded Main Sequence alongside the national science agency, it created a unique way to do that with a model called 'venture science'. As a thesis-driven fund, Main Sequence is trying to solve the same six national challenges with innovation as CSIRO (discussed in Chapter 5). The venture science approach starts by choosing one of those challenges and identifying a global opportunity in line with its market vision, sometimes informed by drawing on CSIRO's science-driven forecasting. The Main Sequence team then assembled the science capability to invent the product for that opportunity, from either CSIRO or a university partner, and began to look for a pathway into market through a leading industry player who would gain significantly from the innovation. Finally, they inject venture investment to create a brand new company.

Venture science investing brings the deep-tech investment expertise of Main Sequence to the table to de-risk, or de-jargon the risk, for corporate investment in deep-tech innovation. Venture science takes the model of venture investing in science from Silicon Valley decades ago, which was all about helping the scientist become an entrepreneur, and now comes at it from the other side to help large, risk-averse industries become more entrepreneurial, reflecting the needs of the Australian system. This aims to pull industry towards investing in science-led innovation and help close the gap across the second Valley of Death (see Figure 2, page xi).

The idea was to create a founding team to share equity in the company in return for bringing the best of each background to the table: the market vision and 'risk' capital injection from VC; the IP, scientific expertise and facilities from research; and the pathway to customer and resources to scale up from industry. Main Sequence would recruit a CEO to work with the founding team on developing and driving the company towards its first milestone, such as developing a demonstrator prototype of the product. As it reaches each milestone and the needs of the company change, the contributions of the founding team are reviewed to ensure the right balance of support as it gains maturity.

As with anything innovative, this couldn't be a one-size-fits-all approach, but there were some core principles needed to maintain the spirit of the Innovation Fund. As a CSIRO initiative, and therefore national benefit-oriented, any new startups had to be additive to existing industries, not set up to disrupt and steal the lunch of existing industries. As I've said throughout this book, I believe this is an inherent value of science-driven innovation. We needed to protect the IP of the founders, avoiding the situation of first-time founders being unable to resist the offers of a big corporate partner with deep pockets. Even if the corporate partner was Australian and a creator of domestic jobs, we wanted the new company to be the job creator, not absorbed into a megalith where jobs would be hard to identify and track, let alone ensure they stayed in Australia. We wanted the startup to be able to access the multidisciplinary depth and breadth of expertise within CSIRO, as well as the rest of the research sector, so it could develop its technology up and down the supply chain without the financial and time restraints of having to recruit a diverse in-house research team. We wanted to use this new model to tackle some of the existing challenges around research–industry partnerships that have seemed intractable for decades: short-term industry investment in innovation, changes in corporate or research leadership, and a lack of common objectives between the research and industry sides of the partnership. These challenges don't seem to arise when venture invests in deep-tech startups, so we wanted to ensure they didn't resurface when a corporate partner was added to the recipe.

One of the CSIRO scientists involved in the first venture science model of company creation, Dr Mary Ann Augustin, published an analysis of the creation of plant-based meat alternative startup v2food in *Global Food Security* journal.[236] She articulated both the benefit of the model to rapidly address challenges in food security as well as to address systemic issues in research investment. Her work captures the collaboration challenges and the potential solutions offered by the venture science model beautifully, and I encourage you to read her full paper in addition to my brief summary:

- *Industry funding*: Business expenditure in R&D is subject to annual funding cycles, changes in company strategy and external market forces, to name just a few factors. Under the venture science model, funding is only required from the corporate partner once the startup reaches agreed milestones, with earlier, higher-risk investment contributed by the venture capitalist to provide more financial security for technology development. The co-development of the startup and product, as well as the equity arrangement, increases corporate buy-in.

- *Executive sponsors*: This challenge applies on both the research and the industry partner side, where a project can be left floundering for executive sponsorship if there is a change in leadership or if the executive sponsor has competing priorities (similar to the example of Wave, discussed in Chapter 3). The appointment of a CEO and a board gives focused, continuous leadership that allows for the longer timeframes technology can take to develop, and results in increased engagement and commitment from the founding team.

- *Part-time teams*: Collaborative research–industry partnerships are rarely the sole focus for participants on either side of the collaboration, which can lead to lack of focus or competing priorities; both sides may prefer different modes of engagement, which can complicate how they engage; and ultimately deadlines can slip on either side for these reasons or others, resulting in disengagement and missed milestones. Appointing a CEO and building a team within the startup enables the creation of dedicated roles, engaging in a unified way and focused on commercial outcomes.

- *Different priorities*: Research organisations want to get maximum impact from their research, while industry partners often look to narrow the remit of the partnership to ensure focus and reduce costs, not seeing broader opportunities. This clash of competing priorities can slow the whole project down. The venture science model gives the industry partner first-mover advantage and delivers the product from concept to market, with the potential for exclusivity depending on the arrangement, as well as a stake in a successful business, while protecting the IP for the research partner.

Mary Ann also noted that co-developing the company from the ground up means each founding team member sees that the company could not have been formed without the value that they each bring, removing the chance of dispute over value that may have been created before other partners joined the venture; this can happen in other company creation models where a startup is spun out of a research organisation and equity and royalties flow back to the parent organisation, or if a founder is replaced with a more 'traditional' CEO further down the road.

Venture science companies are created by a team of seasoned investors, industry leaders and scientists, and taken rapidly to market to reassure risk-averse Australian corporate partners without experience or appetite for investing in slow-burn, deep-tech startups. As Mary Ann articulated, it creates a pressure-cooker environment for a scientist:

> The rapid speed to market with the new innovation model was not without challenges. It required a focus on delivery of innovation with a team more used to delivering scientific outcomes, who had to learn to rapidly prototype products. This required iterations to be developed in response to market signals and the ability to make decisions on research direction to develop 'best bets' with limited information. Constraints of price of ingredients and pragmatic choices to utilise existing production processes were needed to produce a high quality [v2food] burger which also met the consumer's expectation on price, quality and utility. This caused tensions and forced new ways of working at a speed that was uncomfortable and met with occasional resistance. Frequent team sessions and occasional senior management escalation was needed to keep the development on track. Nonetheless, the experience was reported as positive for the scientists involved because of the clear purpose and demonstrable impact of the work.[237]

If the recipe had called for blood, sweat and tears, the team would have had those in abundance as well. Although Mary Ann and the team went through a baptism of fire, it's her final sentence that gives me the greatest joy. Like true missionary CEOs, the team were buoyed through adversity 'because of the clear purpose and demonstrable impact of the work'. Under a venture science model, any company created is going to be harnessing the power of science to solve the greatest challenges, so it's not hard to see why the work was ultimately rewarding and suited to the diverse experience of a scientist CEO.

Aussie precedents for venture science

Although the Main Sequence model for venture science is new, it's not without comparable precedent, and it's the fact that two of Australia's greatest deep-tech company success stories were founded in a similar way that tells me this is a strong recipe for Australia. Former CEO of hearing device company Cochlear, Catherine Livingstone, said there were three core elements to the creation of Cochlear: IP being made available by the university because it was not continuing the research; the research being identified by a public servant as a strong candidate for government commercialisation funding through a public interest grant; and the market vision of Australian entrepreneur Paul Trainor, founder and CEO of medical device company Nucleus Limited, to support Cochlear within Nucleus. Catherine explains:

> Paul Trainor ran Nucleus and said about the Cochlear technology developed by the University of Melbourne: 'Okay, we'll see whether there's any market here,' – starting with the market, not the science – and, 'Okay, we're prepared to lose a bit of money doing this, but we really need that public interest grant and we need that IP on a royalty basis.'
>
> Nucleus had its own financial struggles to manage while funding the shortfall to keep Cochlear going, because the public interest grant wasn't enough on its own. Cochlear paid a royalty on subsequent sales, which more than repaid the public interest grant and, of course, significant royalties were paid to the university.[238]

Catherine took Cochlear through its IPO in 1995 and it's now a $13 billion company. The other example is $31 billion sleep apnoea device company ResMed.

ResMed

Similarly, ResMed had its origins within healthcare company Baxter International through an effort to commercialise technology developed at the University of Sydney. Dr Peter Farrell was Managing Director of the Baxter Centre for Medical Research and Vice President of R&D for Baxter Healthcare, Japan, in the 1980s. He had recruited Chris Lynch, who he studied with at the University of Sydney, when Chris returned to Australia from a stint in the UK. Chris became aware of Professor Colin Sullivan and his team's work to develop the continuous positive airway pressure (CPAP), the first successful, non-invasive treatment for sleep apnoea, which causes accelerated kidney disease and type 2 diabetes, and can shorten life. Peter recalls going with Chris to meet the team:

> I remember sitting down with Colin in his office and he described how his patients wear this machine every night. The Hitachi blower which Sullivan had used was obviously sourced in Japan, and he had a Swiss engineer making bespoke masks in the basement of the medical school. He called in one of his patients, Eddie Merck. I remember it like it was yesterday. Eddie walked in and he had marks on his cheeks from where the mask attached tightly every night to avoid leaking air when he slept. He had necrosis on the bridge of his nose where the mask was digging into him, even though the masks were all bespoke.
>
> I said, 'Eddie, this machine sounds like a freight train!'
>
> He said, 'I moved my bed to abut the garage and I drilled a hole in the wall and put the machine beside the car in the garage.'
>
> I said, 'Good God, Eddie, you use this every night?'

He said, 'I do. Let me tell you what my life was like before CPAP. I'd go to bed for 10 hours, I wouldn't sleep. I'd get up in the morning, go to breakfast, fall asleep in front of my wife. I'd hop into the car and at the first set of traffic lights, I'd nod off. I'd go into work. When I got into the office, I could barely do a tap of work. I just spent the whole time walking around the office trying to keep awake. If I sat in a chair, I'd go into spontaneous REM sleep. I'd drive home the same way I went in. We couldn't go out to dinner with friends, couldn't go to an opera, couldn't go to a movie, I'd just simply nod off. The first night I went on this CPAP, I had my first dream in 15 years. The bottom line is, I would sleep on hot coals if I could get this result.'

I said, 'That's more than interesting.'

It was my colleague at Baxter, Will Pirrie, who told me to look for the 'fatal flaw first'. What is the minimum amount of money and minimum amount of time and effort you need to invest in order for it to work? If it works at the boundary condition, then put the big bucks in and go for it. At this stage, Colin had 100 patients on this treatment and he'd patented it, so I thought my work had been done for me.

I put Baxter's money into it in 1986, but in August 1989, Baxter sold off the division for which we were developing the sleep technology. I said, well, we're going to have to buy this otherwise it will die on the vine. I went to the President of Baxter, Jim Tobin, who later became President and CEO of Boston Scientific, and suggested that Baxter take 30 per cent equity, [ResMed] have 70 per cent and everybody pays for their shares upfront, no special favours or special deals. He said no to equity. So I proposed giving Baxter half a million bucks upfront and five per cent royalty net profit after tax for the next 100 years or whatever.

He said, 'Okay Peter, under one condition. When you're next in Chicago, I want you to buy me a beer.'

And I said, 'Well, wait a minute, Jim. Are we talking import or domestic?'

That's literally how it happened. Within three weeks, we owned it. That was May 1990.

Things limped along for a while but John Plummer came to the rescue with an investment after he had sold his personnel agency to a Swiss company in a multimillion dollar deal. We also had some money from Medtronic in the US.

In June 1995, we did an IPO on NASDAQ for $85 million. If you fast forward to today, we're at about $35 billion in market cap, close to $US4 billion in revenue, we make 20 per cent net income to revenues. We started with 6 people, we now have 8500 in 140 countries.

Drawing parallels with the venture science model, ResMed was founded with IP from an Australian university, external investment through a high net worth individual (where venture could play that role today), and shepherded to market, initially, through the established channels of a large industry partner in Baxter International and then through the expertise of key Baxter leadership members like Dr Peter Farrell, who became ResMed's CEO, and Chris Lynch, who became a ResMed Director.

Both ResMed and Cochlear needed the initial support of established parent companies before they were ready to take off the training wheels and become their own strong companies.

Venture science in action

Over CSIRO's more than a century of solving Australia's greatest challenges through science, our largest and oldest field of research is agriculture – from chasing out the pest of prickly pear in the 1920s to innovating at the cutting edge of genomics in crops in the 2020s. So it is probably no surprise that Main Sequence's first official venture science company, plant-based protein company v2food, as well as another company that was nearly our first venture science model of company creation, carbon emission-abating livestock feed FutureFeed, came out of our agriculture and food research. We set out specifically to create two companies to address our national challenges and give them both a unique advantage from science. This ability for the investors to be part of the founding team, to actually put the company together from a focused market vision, is the hallmark of great venture science, as distinct from a startup coming in to the venture capitalist to pitch their already made deal – I'll admit it's far easier to do when your R&D lab is Australia's national science agency.

Two of Australia's greatest economic opportunities are in renewable energy and space technology, which led to two further venture science companies – hydrogen battery company Endua and satellite communications company Quasar Satellite Technologies. As the following case studies show, they have grown the market and given back to society, while also carving out market-leader roles for themselves and their corporate partners. The industry partners in these deals often also share a common thread of being 'scrappy challenger' companies in their industries, those with their sights set on the top of the industry, not necessarily those who are already at the top with everything to lose, or perhaps they're at the top but have competitors or industry changes nipping at their heels, keeping them innovative. This will be essential company

DNA for forming a new, innovative deep-tech startup, born from an acorn or gumnut.

v2food

The starting point for venture science has to be your market vision and, as a thesis-driven fund, Main Sequence had already identified the six challenge areas it would invest in, aligned with the six challenges CSIRO sets out to solve with science. In the v2food case, the Main Sequence challenge is solving how we feed 10 billion people with the resources of one planet by 2050. CSIRO's Futures and Insights teams – the forecasting arms of CSIRO – had recently found that the Australian market for alternative proteins could be worth $4.1 billion by 2030, with another $2.5 billion in export opportunity by 2030.[239] Once you've decided that alternative protein has to be a massive market because it's got to solve a big problem, then you ask the question: who are the beneficiaries of that? Other than the people you feed, consider people in the market already: who's going to win? In Australia, you quickly recognise fast food as one of the largest industries feeding the country.

Jack Cowin is the founder of Hungry Jack's and CEO of its parent company, Competitive Foods. Jack recalls Hungry Jack's had already started implementing their market vision of selling a plant-based protein burger when the opportunity came up with CSIRO and Main Sequence to make a vastly superior product:

> We're in the meat business and we sell something like 35 000 tons of hamburger patties to 26 different countries. We have our antenna up as to what's going on in that industry around the world. Going back, maybe 4 or 5 years ago, we became aware of Impossible and Beyond, two US entries into this world. They were getting lots of publicity and had a good story to tell that we all have become familiar with.
>
> Having a meat production facility, I thought that it made good sense for us, if this area was going to grow, that this could become another product that we could run through our facility that didn't represent having to go out and build a new plant and everything else that goes with that. I wrote a letter to the CEO of Impossible saying, 'Here's where we are, here's what we do, we've got the capacity to be able to produce your product, this could give you an entry into Australia with existing customers.' That led to no response. Probably this guy was busy doing what he's doing and Australia probably wasn't on his hit list. We then developed our own plant-based product and it wasn't great, but it was a start into in the field.
>
> Not long after that rebuff of no response, I hosted a lunch to discuss quantum computing and met [Main Sequence

partner] Phil Morle. We were talking about what else they were doing and seeing, and I asked him what he thought about this new alternate meat business. Phil said, 'CSIRO has 2500 PhDs all over Australia, we can make anything that is sensible, why don't you let us have a crack at it?' I said, 'Good idea.' That's how v2food was initiated.[240]

There is a team in CSIRO who have been working on the science of protein since the 1970s. They'd invented all sorts of interesting things over the years, but never really found a killer application that could translate their research into massive impact. It turns out they were just waiting to be asked how to make a legume-based protein burger taste as good as – or better than – an Angus beef steak. The challenge for others who have attempted the recipe has always been getting the texture right, but that was a completely trivial problem to the CSIRO team. It took a few attempts before they nailed it – nothing like the years you might usually expect to wait for a science solution. Once we had a custom product, we needed a custom manufacturing solution to make this uniquely textured burger. CSIRO's national research labs and facilities provided the answer – where else would you find a building that's already food-safety qualified to serve as your pilot production facility, and ready to go within a week? – and CSIRO's manufacturing team developed a solution in record time.

It was fascinating to watch it all come together. Each of the experts thought they were doing specialised science and yet no one had ever thought to put their results together in that way before to achieve this amazing result. Having Jack on board to build the concept really focused people's minds; there's nothing like having a customer who is an investor, who's committed and who can get you to revenue quickly by putting your product in their stores and helping you to turn it into a real business.

Jack said partnering with CSIRO and Main Sequence de-risked the investment in many ways, as well as Hungry Jack's making it a lot easier for a startup like v2food to scale-up and reach customers:

> I think we put up $1 million or thereabouts. Main Sequence invested as well. There really wasn't much risk in that we didn't have to build a new plant, we didn't have to invest a lot of money in equipment. If v2food had to build a plant to make these products then it would have been a lot more expensive. We were able to convert what we had in order to produce what we wanted to sell without having to spend a lot of money. So we were somewhat fortunate that those stars aligned.
>
> In addition to the meat processing business that we owned, we also had 450 trial kitchens in the form of the Hungry Jack's business that we could put product in and test. We've got live customers so we'd be able to get a read on where the opportunity was or wasn't. We did that and we started selling the product through our Hungry Jack's facility.

CSIRO were the keepers of technology that we lacked, of being able to develop tastes and things like this. They were able to improve the product that we were currently making. I think what they came up with was better. Getting the product right is important.

Probably more important was when Main Sequence went to the market to raise the $200 million. That was an endorsement that this business could continue to evolve and develop more products and be more competitive. So, I think that endorsement, going to the outside investment in our early stages, I think it gave us more confidence to push on.

The most significant thing that CSIRO's connection to the v2food business was when you go to start to attract outside money, it gives people more faith that this is more than just guys trying to do something as a backyard business and that we have the capacity to be able to evolve and develop new products.[241]

In terms of structuring the deal to create v2food, the negotiation was very much a game of give and take. CSIRO was not prepared to sell the IP, so that was off the table from the outset – the whole idea of Strategy 2020 was to take more risk and earn more reward when our bets paid off. However, we were willing to give Jack an exclusive share of the fast food market through his fast food chain in order to get our idea into market – as Frank Levinson said in Chapter 5, sometimes you have to give something away to realise your best idea. Jack could see the significant value that would bring to his revenue base and he was also smart enough to see that if he tried to build this company within Hungry Jack's, or the parent company of Competitive Foods, he would have to carry the high risk and high complexity, albeit for high reward. So Jack and Main Sequence put both money and some expertise in, while CSIRO brought the science. He knew that if we were successful, he would be the beneficiary because it would feed his supply chain and he knew how big that could be, so he agreed to only having fast food exclusivity, not supermarkets or other hospitality venues. Whether v2food itself was financially successful as an investment became less critical for him, because it had solved his supply chain problem. He was willing to give away the financial gain from exclusivity of IP in return for market exclusivity.

De-risking was always going to be important for a company like Hungry Jack's to invest in innovation and, as Jack sees it, even more important than creating something really innovative:

I've got this thing where I've got to test, test, test, test, test, to be able to verify before I take a risk. That takes longer, puts a lid on some unique Eureka ideas that somebody's going to come up with that are going to blow the doors off, but you're eliminating risk. The more testing you do, the more verification you get on something, the less the risk, and the greater opportunity to be able to succeed.

The objective was not, 'This is going to be a revolutionary new taste', it's not something that's going to be totally new and different. It's, 'How do we make something that tastes like meat?' That was the project. That was the target. We're not trying to revolutionise anything, but what we are doing is we're making a plant-based product rather than something that's going to come from an animal.[242]

And yet, that de-risked investment has created a company that has pioneered a new industry for Australia and created nearly 100 new Australian jobs by the end of 2022. Australia's alternative protein market is projected to grow from nothing to $10 billion in line with one of CSIRO's missions, discussed in Chapter 8.

Main Sequence kicked off preliminary conversations in mid-2018. By October that year, we had a founding CEO identified and in January 2019 v2food became operational, named as a shorthand for 'version two' of meat. In October 2019, the first v2food 'meat' patties went on sale in Hungry Jack's. The following month, v2food raised A$35 million in series A funding, followed by A$77 million in series B funding in October 2020 and another A$72 million in a follow-up series B+ round in August 2021, bringing its total to over A$184 million raised in less than 2 years. Shortly after the first products landed in Hungry Jack's stores, a range including burger patties, mince and sausages, all tasting like beef, began progressively being stocked at major supermarkets around Australia, with plans underway to realise the goal of exporting into Asia. v2food has also invested capital into building its own production facility in Victoria, creating infrastructure for future growth.

If you go back to the thinking that initially took us to Jack Cowin, you could also ask: who else in the supply chain would have had a stake in this development? When Main Sequence was ideating v2food, we thought about approaching legume growers, as we knew there were some quite large consortiums who wouldn't mind breaking into a new market. We decided we'd do that after the company was up and going, and they'd be chasing us rather than us chasing them, but if we'd been in the US, we might have thought differently about that because there's less risk aversion in their agribusiness industry.

FutureFeed

The company that could have been our first venture science deal was FutureFeed. CSIRO scientist Dr Rob Kinley showed me the original idea on my first visit to our Townsville site and it was impossible not to support it. He and his colleagues from James Cook University had found very promising early results in trials to add a seaweed supplement to livestock feed to reduce cattle methane emissions. We funded Rob and the James Cook University researchers to take the idea for

FutureFeed through CSIRO's first round of the ON accelerator in 2015. CSIRO could see the value and provided the funding to get the idea off the ground while it looked for investors.

I can't tell you how many times I pulled a bag of FutureFeed seaweed out of my pocket to show potential investors or customers – even having to explain what it was to suspicious officials at various airports. There was immense scepticism about this idea and little support initially. There was an understandable concern that this would benefit other countries more than Australia, as the use of livestock feed in feedlots is less common in Australia where we favour grazing in fields. Eventually a combination of developing more ways to feed cattle FutureFeed and recognition that climate change is a global problem, not just an Australian one, won the day.

FutureFeed was finally launched as a company in August 2020, with CSIRO, James Cook University and Meat and Livestock Australia as the research partners and IP holders, and investment from a consortium made up of Australia's largest supermarket chain Woolworths, grain logistics company GrainCorp, agricultural investor Harvest Road, and an additional joint venture between real estate investor AGP Sustainable Real Assets and agriculture accelerator Sparklabs Cultiv8. FutureFeed now licenses its products to existing feed suppliers, with licensees in Australia, New Zealand, the US and Europe so far.[243]

Woolworths is not going to sell cattle feed, but it does sell meat and worry about environmental, social and governance (ESG) considerations. So a little bit like Hungry Jack's with v2food, there will be a flow on benefit to Woolworths' supply chain through the uniqueness of being able to sell 'high quality, low-carbon beef', whether FutureFeed is financially successful or not. Historically we haven't often seen Australian companies do that, but I hope this is a sign of the tide turning.

We put in the first few million to make FutureFeed and v2food real, but it was industry – companies like Woolworths and Hungry Jack's – that came in afterwards to invest in turning them into real Australian businesses, something that can scale and employ Australians.

During COVID-19 lockdowns, Qantas CEO Alan Joyce hosted a virtual cook-along with celebrity chef and Qantas partner Neil Perry, cooking a beef steak from FutureFeed-fed Angus cattle and a v2food meat product. My daughter Jessica helped me cook both in our family kitchen and assured me that, as a vegetarian, v2food isn't competing with real meat. As I suspect all the investors know, both FutureFeed and v2food are growing a more sustainable and profitable protein industry with more options to feed our growing global population.

Endua

Energy is another one of CSIRO and Main Sequence's shared challenges, with Main Sequence focusing specifically on 'decarbonising the planet'.[244] The proposal for developing a hydrogen generation and storage solution was initially proposed as part of CSIRO's Future Science Platforms, a program for breakthrough new research that needs further exploration before it's ready for commercial applications. The technology would be a solution to support renewable take-up in remote applications where transport of liquid fuel is costly, carbon intensive and logistically challenging.

This is where the beauty of having the ON accelerator, Main Sequence and a program of CSIRO missions (as discussed in Chapters 6 and 8) originating from one organisation like CSIRO comes into play. The fund partners invest in startups aligned with the challenges CSIRO has set out to solve, so they were already looking for hydrogen startups when they heard about this research proposal. They identified an opportunity to align with petroleum company Ampol's Future Energy and Decarbonisation Strategy and took the researchers to meet with Ampol, who supply numerous regional and remote communities.[245]

Announced in June 2021,[246] the hydrogen startup Endua was formed with an initial $5 million from Main Sequence for CSIRO to continue developing the hydrogen technology. As Australia's largest fuel network, Ampol took a stake that involves bringing industry and customer knowledge to test and commercialise the technology, with more investment from them expected as milestones are met. Ampol will feed the increasing need for low-carbon, off-grid alternative energy solutions among its base of about 80 000 business customers nationally. Sales will initially focus on the off-grid diesel generator market, which accounts for $1.5 billion of diesel and 200 000 tonnes of carbon emissions every year.

Endua's founding CEO, Dr Paul Sernia, studied electrical engineering and computer systems engineering before moving into business roles in the electric vehicle industry, including also being a co-founder of electric car charging station company Tritium, discussed earlier in this chapter. As a scientist CEO himself, Paul is supported by CSIRO Principal Research Scientist Dr Sarbjit Giddey as Chief Scientist. They also have expertise from Ampol and Main Sequence on their board.[247]

Quasar Satellite Technologies

CSIRO has been a leader in radio astronomy and spacecraft communications for more than 60 years, from supporting the Moon landing in 1969 to inventing and delivering the phased array feeds in Australia's newest radio telescope, ASKAP, in Western Australia in 2013. CSIRO's technology breakthrough enabled the world to connect without wires using fast wi-fi and now its technology will help connect satellites using ground-breaking phased array technology.

With the satellite ground communications market worth US$130 billion, Quasar Satellite Technologies was created to commercialise the breakthrough science developed by CSIRO for radio telescopes, which will help reduce congestion in data traffic from space as more and more devices join the growing 'internet of things' movement.[248] Quasar's first funding round was backed by $12 million in capital, technology and industry expertise from CSIRO, Main Sequence, the Office of the NSW Chief Scientist & Engineer and Australian companies Vocus, Saber Astronautics, Fleet Space Technologies and Clearbox Systems.

Quasar's leadership team is also strong on scientist CEOs, with computer engineer Phil Ridley as CEO, former CEO of NICTA and co-founder of wi-fi company Radiata Professor Dave Skellern as Chair, and former CSIRO astrophysicist and commercialisation specialist Dr Ilana Feain on the board, together with expertise from Main Sequence and Clearbox.[249]

Australia needs its big businesses to engage with deep-tech innovation if we are going to create a thriving innovation ecosystem. Silicon Valley and Israel have strong cornerstone companies, but Australia will not be innovative or successful by playing catch-up on their terms. The venture science model can help us leapfrog the time it will take for our fledgling deep-tech startups to become business behemoths. It also creates space for scientist CEOs in smaller companies; manages board risk with deep-tech VC members; and seeds innovation into our existing industries by supercharging corporates that have challenger DNA. It's also my hope that as more of corporate Australia leans into this space, it will challenge the risk aversion of Australian venture capitalists, who will lean in further to get their share of more of these deals.

Venture science is important for Australian companies because it gives them a safer, more collaborative way to consider risky stuff. If you can make them part of the journey from the beginning and add VC investment at the earliest, highest-risk stages, you increase the probability that they will jump in, and then you have a powerful force in the innovation ecosystem.

Part 3: How to transform the system

Innovation is not a linear exercise of transferring knowledge from a lab bench to a customer. It's a series of iterative loops, steps forward and backwards, and it takes a thriving and layered ecosystem to support scientists and their ideas to leap safely across the Valley of Death. Part 1 focused on how to equip our scientists for the leap while Part 2 looked at opportunities for large, established companies to lean in from the other side of the Valley and reap economic rewards along the way. Part 3 now builds on all of these elements to consider what it will take for Australia to grow the kind of system that will make leaping across the Valley of Death not only easier for scientists, but much more common. In many ways CSIRO is an analogue of Australia – it operates across all the key market verticals of Australia, and it has gone through a long, slow 20-year decline despite, or perhaps because of, 30 years of uninterrupted economic growth. But CSIRO managed to reinvent itself before the catalyst of COVID-19 when Australia again turned to science, and has delivered exceptional returns. CSIRO regularly engages external, independent auditors to assess the return on investment that Australian taxpayers receive from CSIRO's research. Over CSIRO's current strategy period (2014-2022), auditors found the environmental, social and economic value that CSIRO delivered to Australia grew from $5 billion in 2014 to $10.2 billion in 2022 from an organisational operating budget of $1 billion in 2014 and $1.4 billion in 2022.[250] CSIRO can teach us how to transform Australia's system.

What can we learn from CSIRO? With just 5 years of effort, Australia could become 10th in the world for innovation to match our Research and Development WIPO rating (discussed in Chapter 1); Australia could increase the diversity of voices represented in leadership, particularly in gender and Indigenous representation; and we could grow our equity portfolio of tech stocks tenfold to replace the economic pillar we currently get from exporting fossil fuels.

You can cut through the complexity of driving a system-wide change by finding a common motivator: there must be economic rewards for industry participants, job creation and solutions to policy challenges for government, and opportunities for research excellence and real-world impact for scientists – and, ideally, these collaborative efforts will be engaging and inspirational for the next generation of leaders. It sounds like a tall order, but it's not without precedent – how else could humanity have made it to the Moon? But this time, we don't need to leave the planet, as there are plenty of challenges to solve on Earth that can simultaneously help us solve our innovation challenge and, in so doing, lay the foundations for collaboration that can overcome any challenges that arise in the future.

8

System on a mission

Science can be many things to many people: it can solve problems, grow wealth, protect our environment, save lives, fight off a pandemic, cross borders to enable diplomacy and inspire future generations. Sometimes it can do all those things at once. Science can be performed by individuals or small teams, but it has far greater impact when it unites large coalitions of diverse groups around a single goal. For science to have this level of impact in a relatively small nation like Australia, we need a way to centralise our efforts and motivate participation. This chapter considers ways to engage the full scale of the innovation system – including research, businesses of all sizes across all industries, government, investors and the community – around common deep-tech goals. In particular, it looks at the legacy of missions like solving the prickly pear invasion, going to the Moon and eradicating polio for a model of modern missions that will solve our greatest challenges today and inspire the next generation of STEM-engaged leaders.

Missions

CSIRO's researchers have worked with partners to develop solutions from science for their problems for over a century, but some of those goals have been bigger, bolder and further reaching in their ambitions than others. Improving a company's product or process is high impact for that company and, ideally, creates more jobs and economic prosperity, but it doesn't necessarily make life better for every Australian. When researchers, businesses, governments and the public become excited and engaged with solving a national challenge, over time we've used words like 'mission' to describe them.

One hundred years ago, CSIRO's first mission was working with partners in the New South Wales, Queensland and federal governments, as well as farmers around the country, to tackle the scourge of prickly pear, an invasive cactus that had spread across more than 24 million hectares of Australia's farming land by the mid-1920s. Australian researchers searched the planet for the plant's natural predators, eventually identifying the Argentinian moth *Cactoblastis cactorum* and releasing it into the Australian landscape in 1926. Within a decade, the prickly pear was almost completely gone.

A couple of decades later, physicists at CSIRO reinvented radar to defend Darwin in World War II,[251] and a couple of decades after that the same group built the CSIRO radio telescopes that were part of the mission to put humans on the Moon, with the first images from that historic voyage reaching the Earth via our Canberra Deep Space Communications Centre in July 1969. Announced by US President John F Kennedy in 1961, the ambition to go to the Moon was not a goal that could be achieved by the US's National Aeronautics and Space Administration (NASA) alone. Not only did it take an international network of space agencies and infrastructure to support their efforts, but it also took the US's best researchers in manufacturing, energy, textiles, food, biology and a range of other fields coming together to take on this never-before achieved feat. As has been well-documented, just the process of preparing to go to the Moon spawned new inventions that made life on Earth better.

Internationally, there is an increasing focus on challenge- or mission-based innovation, epitomised by the German Hightech-Strategie, the Japanese Moonshot Research and Development Programme and the UK's Grand Challenges.[252] In 2013, then-Professor of Economics at Sussex University (now at University College London) Mariana Mazzucato published *The Entrepreneurial State: Debunking Public vs Private Sector Myths*, lauding the critical role of public investment in science in driving the private sector's innovative products. Most famously, she details all the ways publicly funded research developed different elements of the iPhone, like DARPA's invention of the internet and of the voice recognition technology that powers Siri; the US military inventing GPS; and touchscreen technology being developed by a spin-out from the University of Delaware with grants from the NSF and CIA.[253] Building on this seminal work, she became a champion for mission-oriented research models to drive public–private innovation partnerships and for developing a framework of broad 'grand challenges', similar to CSIRO's challenges, which should be solved by goal-oriented and time-bound collaborative 'missions'.[254] She took on the role of Special Advisor for Mission Driven Science and Innovation to European Union (EU) Commissioner Carlos Moedas[255] and, under her influence, the €100 billion Horizon Europe fund, announced in 2021, is investing €4.5 billion into five 'missions' in areas of climate change, cancer, oceans, smart cities, and soil and food.[256]

I met Mariana when she came to Australia in 2018 to talk about the model of missions she was advocating for the EU to adopt. We discussed how to design a formal program of missions in Australia to accommodate the differences between what would work for the EU versus what Australia needed. A few things leapt out at me.

The first was that Australia doesn't have a great mission-like breakthrough to celebrate the way other countries do. *Entrepreneurial State* was full of examples of the importance of government-funded research to drive national innovation performance. She talked about Sweden inventing Bluetooth through their telco Ericsson, which had been saved from bankruptcy decades earlier through government and private investment; and she discussed the ongoing financial returns the Finnish government received from retaining equity from their early stage investment in their telco, Nokia.[257] She said we were missing an iconic company and invention that we could point to with pride. As mentioned in Chapter 1, I asked her if she'd heard of wi-fi and she said, 'Sure.' I said, 'But you've never heard that Australia invented wi-fi?' She was stunned. She said, 'I've been here in Australia for a week and you're the first person to point that out to me.' It really shocked Mariana and she said she'd put it into the next update of her book because she was looking for an example for Australia.

The second thing that struck me about her core thesis was that there are certain things, like market failures, where government has to invest, otherwise they can't be solved. A mission like going to the Moon is a classic example – we never would have gone to the Moon if the US government hadn't thrown billions of dollars at the problem. I agreed there are some things where government has to intervene, but I told her I was more interested in other things that are just as important and just as mission-like, but that can be funded by industry or by venture. She was sceptical, but I've always believed the key to solving our innovation dilemma will be an innovative Australian approach, so her scepticism seemed like a good sign to me. Mariana has done a great job of marketing the concept of missions globally and doing something that every entrepreneur has always done: create a bold vision of the future, use it to inspire people to join you who wouldn't otherwise and take them with you on that journey to market. It's still all about having a market vision to drive you.

Special Innovation Advisor at the University of Technology, Sydney, Emeritus Professor Roy Green points to Australia's history of innovation interventions to suggest we have always taken a 'linear commercialisation' approach to innovation, where you take an idea straight from the lab to the market by 'mining IP in universities' as though such ideas are just sitting on benches waiting to be commercialised. Instead, he advocates shifting to an 'innovation ecosystems' approach, where you create longevity and constant interaction between key players in an area of research and 'ideas emerge from constant interaction'. He said, 'You start on a project and you might end up in a completely different place if you've got exposure to a much wider circle of innovators, whether they're in industry or in universities, especially if they aggregated in one place, as they are for the [UK] Catapults now and Manufacturing USA Institutes.'[258]

Similarly, Main Sequence partner and ON mentor (discussed in Chapter 6) Phil Morle said the next challenge for Australian innovation is to move beyond successful programs and linear interactions and into successful systems and networked interactions:

> Whilst Australian universities have always had a solid practice of 'tech transfer', this has been a shallow approach to commercialisation. In these models, IP developed inside a university is licensed or assigned to external firms to commercially exploit. Universities have not captured the value and have not maintained a connection with the IP that could drive the impact further.
>
> There have been huge leaps in this practice since 2015 with a number of universities developing a practice for spin outs and a culture of 'permeable walls' for startups to work within a campus, connecting research with commercial innovation. Main Sequence arrived as a venture fund custom-built for this context and created impactful companies quickly in partnership with universities. For example, 5 years after Main Sequence's foundation, ANU [Australian National University] has active collaborations with 10 Main Sequence portfolio companies.
>
> ON existed to deliver accelerated mindset shifts and knowledge of new tools. Now we need to move this to systematic company creation as the base skills exist sufficiently and ongoing training has now spread to the edges. The universities are teaching themselves. I'd like to find a real connection (one that can be operationalised) that connects longer-term industry builds across research, startups, government and industry. No one of these groups can do it alone and the current models are still too fragmented. CSIRO's missions are an example of how this could happen. Let's bring people together more and build the future instead of squabbling in our bubbles about what is not working.[259]

Those interactive, multidisciplinary ecosystems aligned around a challenge are exactly what missions are designed to be.

National missions for Australia

CSIRO has been at the heart of Australia's innovation system for a century and I firmly believe it has a pivotal role to play in transforming our system, but I also know one CSIRO program alone cannot solve all of Australia's – or the world's – challenges. Nevertheless, I knew we needed a prototype within CSIRO to start the change, so we began planning for a uniquely Australian model of missions that could catalyse the system. We started with CSIRO's market vision – the six national challenges discussed in Chapter 5 – to guide the missions we would focus on.

The missions model requires a few basic elements, like being co-developed through deep collaboration with partners in industry, government and research and setting out to solve bold, visionary and inspirational challenges with specific, time-bound, measurable goals. To that recipe, we added a few uniquely Australian elements that both leverage our strengths and tackle our innovation barriers (as discussed in Chapter 1), as I discussed with Mariana all those years ago. First, Australian missions are more strongly funded by industry than government, a reversal of the EU model. Second, they should drive cultural change by embracing collaborative networks instead of linear or hierarchical ways of working, as well as running towards risks in an effort to fast fail – or fast succeed. This helps narrow the gap across the first and second Valleys of Death (see Figures 1 and 2, pages x and xi).

Because they are designed to tackle our barriers, missions operate in a way that is counterintuitive to our current innovation system. They are trying to force a paradigm shift that we believe Australia is ready for. In designing a missions program that can effect system-wide change, it's my hope we'll dismantle some of the myths around the likes of Stanford University being held up as the paragon of university commercialisation, when in fact it is the thriving innovation ecosystem of the surrounding Silicon Valley that pulls tech off the benches and thrusts it into the market. Missions create a system-wide environment that embeds innovation at every touchpoint, creating the strongest chance Australia has had to turn around its innovation performance to date.

Industry-led missions

Australian businesses have not historically been strong investors in R&D, as discussed in Chapter 1, so when you approach them to invest in something that hasn't even been invented yet, the challenge to get them on board is significantly tougher. It's not common in Australia to go to an industry customer and say, we want to go on a mission together to solve this problem. In Chapter 5 I talked about how we worked with businesses to develop their market vision, co-developing a vision for the future and creating an opportunity to apply research and technology to try and tackle their roadblocks. You've got to understand both that customer and the market really well. Your market vision has to be strong and you need to have clear knowledge of what is the biggest roadblock in that market and, roughly, how you're going to solve it. Just knowing what that roadblock is often surprises customers, because they probably don't know it themselves, and the act of unpacking how you identified it is a great way to bring you and customers together. It changes the relationship from customer and vendor to partners who are both on a mission to try to crack this problem – but you only get to that point by convincing them that really is the problem and that you understand their market. Suddenly, you're in the tent together and you're really focused on the same problem. Human nature has shown us that people, particularly scientists and engineers, love to solve problems – and co-creation is a key element to a successful mission.

For example, we have a number of research initiatives across CSIRO that, when brought together, could make a significant contribution towards ending plastic waste. We talked to industry to understand how it thinks about plastic waste and narrowed down to where science can make the biggest difference. It turns out that a lot of plastic waste can be tackled through behavioural change and regulation without needing science and technology to reduce it, but you can really shift the needle by redesigning plastic itself for certain uses. We worked out those specific intervention points with partners in industry to co-create a mission focusing on three key bottlenecks: changing the way we make, use and recycle plastics; supporting a sustainable plastics circular economy; and revolutionising packaging and waste systems.[260] Every other part of the challenge was already being covered by industry, but these three areas couldn't be done without science.

Some of our customers have seen us as so intrinsically embedded in the missions program that we've been part of their internal processes. For example, in co-developing missions with the National Australia Bank (NAB), we've been invited to the meetings where our partners within NAB have been updating their executive team on progress, critiquing the program and making decisions about their next commitments. Getting to the point where our partners invited us into the room to be part of the process and seeing their leaders interrogating the mission to make sure it was really going to work was a great sign of success. That process is not an adversarial one. You're not in the room for the leaders to decide whether or not they're going to do it: they've already committed to the mission. It's also a sign of a great partnership when a company is already thinking in that collaborative way with their partners.

If the company is already experimenting with 'agile' methodologies aligned with the Lean Launchpad way of working, which most modern companies are, it'll get excited when it hears the mission language, because missions operate in a series of 'sprints' to reach milestones, and often those milestones are needed to draw down the investment from our partners. Just like in a startup with different funding rounds, there are natural break points through these sprints where missions may fail in that process. Sprints to earn investment milestones are a way of handling the inherent risk aversion in Australian board and c-suite culture. For example, through working more closely with banks, like NAB, who are continually tightening their governance after the banking royal commission, we've found ways to de-risk their decision points. They need to know how much capital they're committing, what the specific end goal is and, most importantly, the intermediate goals that will tell them whether or not we're on track or if there's a problem – the 'canary in the coal mine' goals. So the process has a fairly light touch, but we make sure that by the time we get to that 'canary in the coal mine' goal there are no surprises because they've had transparency all the way along. When you're headed towards a highly ambitious goal and you're trying not to wind back your ambition to manage your risk, you will surely miss some of your goalposts. That's why transparency is really important, so everyone understands how the goal was missed. It means they can make a rational decision about pushing forward because they've seen enough to know that things are working and agree to set a different goal and sprint to it to test assumptions about progress.

We've also been working with banks on missions related to how they assess risk when it comes to extreme weather and climate change. Other missions are being developed in response to market disruptions that pose an existential threat to major companies, like the net zero emissions threat facing Australia's mining and steel-making companies.

Leading with industry has meant our missions have brought in much more significant investment to CSIRO than usual. CSIRO's missions are clearly a very different funding model to 'business as usual' for us and they do look, feel and operate much more like a startup. They've brought in much higher co-investment ratios from industry and they've attracted government through that. Certainly the missions program that CSIRO runs is completely different to the European one, and when I caught up with Professor Mariana Mazzucato again, she was surprised that CSIRO's missions weren't exclusively government-led. I think it's an area where Australia is innovating in a way that works for our ecosystem.

That notion of priorities, missions, market – it's all the same idea about starting with the problem you want to solve and not the science that you want to do; market pull instead of science push.

Microsoft Australia and New Zealand's Managing Director Steven Worrall said that, globally, even defence departments are realising it makes sense to partner with industry on tech innovation, because government can no longer keep up with the pace of change, regardless of their budgets:

> There's a reason why so many defence departments around the world, and militaries more broadly, are looking to the commercial providers of hyperscale cloud computing and AI capabilities. They've worked out that our technology has now surpassed theirs, and their capacity to invest in the technology as an independent or sovereign entity to build military, offensive capability has been passed by companies like Microsoft, AWS and Google. This presents a clear signal about the shifting nature of how technology is being created and where the forefront of that technology will be.[261]

The risk we've got to manage is that if we're going to depend on industry first, we can't let that bias us to be more incremental or more near term in our challenges. The reason we can get away with taking that risk is that, at the moment, there's so much disruption in the world and there's so much desperation for dealing with issues like climate change that there is appetite for ambitious goals.

The size of Australia's economy means we'll never have the multi-billion-dollar European Sunrise missions fund, or the scale of funding invested in the US model, but I think the fact we haven't has made us a bit leaner, hungrier and more innovative in how we go after these things. We're going to have to be better than other countries because we don't have as much money or as many resources to throw at the problem, simply because we are a smaller nation with a smaller national budget. That should make us more innovative, give us a more unique outcome and absolutely make us more focused. By necessity, we'll only pick a small number of very important things to do that are uniquely important for Australia and that will serve in our favour, like the dandelion analogy in Chapter 1.

Culture change

Missions, to me, are just startups with more people. They leverage the power of networks by aligning people around common goals and embrace a 'fast fail' approach that will tell them if they're on the right track or not. These are all important levers for innovation but, given collaboration in our current innovation ecosystem is motivated by funding and financial success, I'll discuss these elements through that lens.

Missions help us pool and grow funding

Collaboration makes it hard to get recognition that secures you support, whether that's advertising your successes to international students, industry partners, or anywhere else. It can be hard to get recognition when you are one organisation in a cluster of many. I don't think scientists are naturally ego-driven, but they are passionate about their work, so it's critical every partner that's made it happen is included in sharing the excitement. Leaving partners out has a far more significant chilling effect on collaboration than it does in more well-funded innovation systems.

Missions offer a different financial model to try and overcome this legacy. A mission will spawn companies and IP; it is the very definition of a growth-oriented program. From the start, the partners have to talk about how they're going to share this new value. Coming into these conversations requires partners to leave their finite-pie mindset behind, because you can't work with people like that on a mission or a startup. If they see the funding or opportunity pie as finite, then they're always going to be trying to eat each other's slice. Some organisations have very valid concerns about protecting their IP, but CSIRO and others have been writing contracts to manage IP for a long time, so while these concerns are valid, they're not insurmountable. We need to get researchers and industry to realise that, if they both take a bit of risk together, we can actually create a new market, one where we don't have to fight over a share and where everyone who is on the mission can win. Taking the time to talk about that at the beginning of the mission so people understand how they're going to play together for mutual benefit is key to any collaboration.

Missions help us set forward-looking priorities to consolidate that funding

Australia being smaller than the EU or US doesn't mean we shouldn't undertake expensive, resource-intensive missions; instead it means missions can help us identify priorities by focusing on areas where we already have existing talents and interest. Special Innovation Advisor at the University of Technology, Sydney, Emeritus Professor Roy Green emphasises the importance of pairing missions with national foresighting work, like CSIRO performed with the *Australian National Outlook*[262] to align and focus us:

> It's essential that Australia takes a mission approach to where we want to be as part of a foresighting exercise. The foresighting should understand what our capabilities are, what the opportunities are, where industries generally are going, where the areas of specialisation are that we can have competitive advantage in. But at the same time, there's no point in having all that unless we know what the problems are that we want to solve.
>
> That's where the missions become important. The missions on their own don't do much, they have to be backed up by the capability, but if you do it as one big foresight exercise as the Finns do, and a few other places – and CSIRO has approached that to some extent with the *Australian National Outlook* – then I think we've got the makings of a strategy and a structure, or something that will structure people's thinking about what the challenges are.[263]

Further, Soprano Design founder and CEO Dr Richard Favero suggests that core missions would give a useful signal to business about procurement opportunities:

> The missions should identify the big problems and then distribute funding that boosts the Australian economy by buying those research solutions, it would tell the research community where to line up and push out and do the work. It's not that hard to lay out a plan like that. More of that would actually give better outcomes for the investors and for the entrepreneurs who want to back an idea to solve something.[264]

The interesting point of difference in Richard's proposal is a focus on procurement instead of research funding – it would send a signal that investment will come in after the hard work of commercialisation is done to buy the end-products and create a market. It's investing in outputs, rather than inputs.

Picking winners is actually not that hard, even though we've spent decades shying away from it. There is underlying consensus in the system about the priorities. When CSIRO developed its six national challenges as part of reframing our market vision for our people (discussed in Chapter 5), we also started to socialise them externally, laying the groundwork to start developing missions that would solve these challenges. Everyone we spoke to in state and federal governments and universities was developing a 'challenges and missions' framework, but nobody wanted to admit to having the same challenges: they each wanted to be different and unique. When we mapped all of them out, there was alignment across the system – they had all just used different words to stand out from each other. That reinforced my belief that we had picked the right challenges, but also underlined how hard it was going to be to work collaboratively.

Geography will be a factor in helping us set those priorities. CSIRO is co-located with experts in resources in Perth, biomedical manufacturing in Melbourne, quantum and robotics in Sydney and space technologies in Adelaide, just to name a few areas. But regardless of where we have centres of excellence in Australia, every state government and many universities wanted to lead missions in quantum, biomedical manufacturing, space, and numerous other areas, but none of them were speaking to each other. This behaviour, in state governments especially, pre-dates my time at CSIRO; I saw it when I was in venture capital in the US in the late 1990s and 2000s as part of the expat tour meeting with state governments to talk about investing in deep-tech, as discussed in Chapter 2.

Special Innovation Advisor at the University of Technology, Sydney, Emeritus Professor Roy Green believes there's a location-specific element to innovation success, and it's one of the reasons he thinks the CRCs haven't been as successful as they could have been:

> The problem with the CRCs is that they don't have longevity and they are not in one space, they are disparate universities distributed across Australia together with companies. Some people never actually get a chance to meet. It's all done by other forms of communication. They might have a conference every now and again, and the people involved in a project might work together but then others in that CRC would never meet them or hear what they're doing. The problem is the CRCs are just very loose structures. Sometimes that can work to its benefit, but they miss out on that wonderful, serendipitous experience of being in a physical space. We've noticed at Tonsley [co-located with Flinders University in South Australia], and you would see that at Clayton [co-located with CSIRO and Monash University in Victoria] to some degree, in Tech Lab in Sydney, the ARM [Advanced Robotics for Manufacturing] Hub up in Queensland – people just like coming there. Companies just turn up with problems to solve. It isn't that they're putting in a submission for a funded program, they're just there to have a chat and something will grow from it.[265]

I think most states know what they're good at and certainly CSIRO has a thesis about which state is going to be good in which area as part of our market vision for Australia. Missions lend themselves to that focus. If we treated states more like startups coming to a venture capitalist for funding and asked them what the one thing is that they're going to be really good at, we'd probably do a better job of growing out of both the pandemic and the related economic malaise into real, differentiated success. Federal–state partnerships that invest in missions and give each state a leg up in a particular area would be a powerful signal to send to other players in the system, from universities and business right through to CSIRO.

Missions create networks without restructures

When we started developing ideas for missions within CSIRO, we hoped their highly cross-disciplinary nature would lead to intra-organisational collaboration that could effect organisational and cultural change without having to formally restructure anyone. We built in a requirement that a mission couldn't be developed within just one part of CSIRO – it had to bring together people from different parts of the organisation. We'd done this with a few other funding initiatives, like our Future Science Platforms, into which, as of 2022, we've invested $425 million to create 20 longer-term, exploratory fields of research,[266] made possible with returns from equity invested in CSIRO spin-outs and startups. These cross-organisational initiatives have built confidence with our people that we use investment drivers to fuel collaboration rather than restructure them.

Organisational structure is a blunt instrument. In general, big enterprises such as CSIRO use restructure as the tool to drive any organisational change and it's on a hair trigger. Whenever you want to do something differently, you restructure. But if you want to be a truly agile organisation, you need to figure out how to transcend your structure so that you can organise a team quickly with whatever skills you need to go after any problem, be it a mission problem or otherwise. Ultimately, I would love CSIRO to be exactly that, where no one cares about the structure because we operate by organising the work around the problem – through our missions, priorities, challenges and the problems we're trying to solve.

That structure is very much a contemporary digital or software or internet company structure – a holacracy – something very fluid so companies can be more agile. It's much, much harder to have that fluidity in an organisation that deals mostly with hardware and particularly science that has to be invented before it can be done. Admittedly we did a restructure when I first arrived, but in 8 years that's the only major one we've done and so it does feel like we're starting to transcend structure. Startups don't care about structure or titles because there usually aren't enough people to have much of a structure in the first place. It really is interesting how many startup 'rules' or tricks work in a big organisation, but I think many management teams of big companies find a comparison with a startup to be terrifying. They view a startup as like managing chaos, but that is also the process of science, invention and creation. It is, by necessity, quite random and chaotic. And CSIRO has been pretty good at putting systems around chaos to try and nudge it in the right direction without crushing it.

While missions were being developed by this groundswell of collaboration and transcending structure, we did have to spend some time working through the impact of this new program on existing programs. Ultimately though, missions began to bring in more sustainable revenue than earlier approaches had done – longer-term revenue from a wider pool of partners – and began to be accepted. Missions are proof that you get exponential gains from a networked (rather than hierarchical or linear) approach to innovation – it's an entrepreneur's version of Metcalfe's law of networks for telecommunications. CSIRO's business model had been linear for 100 years and that hamster wheel of only ever growing our revenue as fast we were hiring was exhausting for everyone, so the notion of revenue that could grow non-linearly with headcount was pretty compelling. We gave people involved in missions permission to act like they were managing startups, approaching the people from any part of CSIRO and any business, research organisation or government department that they needed to form their agile, single-minded team.

Benefits flow from having missions developed and owned by lots of different organisations across the innovation system. Those benefits are similar to those that corporates believe flow from a model where a single leader drives a key change program, only to step back from it after a time and leave the rest of the organisation to step up and own the initiative. For example, a company may want to bring in a Chief Transformation Officer, a Chief Innovation Officer or a Chief Diversity Officer initially to scope out the program, identify what needs to be done and get some early runs on the board. Indeed, often boards think this is necessary and evidence of their commitment to change. Pretty quickly though, you want that leader to be building a network of champions who are going to embed transformation, innovation, diversity or whatever organisational shift you want into their business as usual, otherwise it will never truly become part of organisational culture and just live in the silo of the responsible leader. Ideally, you want that person to come in for a finite period of time before the change is co-owned across the leadership team and ingrained deeply into all parts of the business.

I think the same is true of missions. CSIRO and others who have started forming missions play an important role in getting test cases up and building models for co-development, but they have to be owned by many across the system for it to lead to system-wide change. When you remove competition for funding and competition for credit you begin to see behavioural changes, like during Australia's early COVID-19 response where we saw the benefits of collaboration and all reaped the rewards. We will never look back at our response and pinpoint a single university, agency or government department that led our nation's strong performance, but we know that our response was an Australian one, not an importing of solutions from overseas. We had competitive vaccine candidates in the international field, we rebuilt Australian manufacturing in response to the international supply chain breaking down, and our own healthcare system supported patients at home and in hospitals, as well as driving world-class vaccination rates. That's the kind of national pride we want to be able to take in our future energy mix, our sought-after food exports, our resilience in the face of droughts, fires and floods, and many other challenges. We can do all of those things if we give up the idea that a single organisation can do it and earn credit for it alone.

Admittedly, the counterargument is that if there is no single person or organisation responsible for the program, there is no accountability. That is definitely a challenge and only works if you've built a network of leaders or champions who truly believe in the importance of the change and see the benefit to themselves of working that way. So in that corporate example, you need to have demonstrated that digital transformation has saved time or money, or has simplified things, or that diversity has brought new and valuable voices to the table that led to better decision-making, so there's a reason to continue with those changed behaviours above and beyond that executive sponsor leading the work. For missions, that could mean longer-term, more sustainable funding models for some partners, it could mean a pipeline of startups that deliver returns on IP, or it could mean efficiencies in business models as collaboration creates new opportunities to share resourcing of common requirements across parties. For our business unit leaders it was realising, just 2 years in, that we could deliver what used to cost us $250 million for $200 million. In a commercial company that would have meant more profit, but for CSIRO it meant more time to think and create the next 'Horizon Three' science. One example that comes to mind is the groundwork CSIRO did ahead of launching its drought mission that led to the creation of spin-out company Digital Agriculture Services (DAS),[267] which earned us a fair bit of credibility about what we could do in this space.

A successful model for missions is not that CSIRO runs them all (as much as we would love that) but that they crop up across the country, led by many and varied organisations that bring together many partners around a truly national, focused effort that gives us critical mass without duplication. A big victory for me, and actually an internal KPI we've set at CSIRO, would be for us to give a mission to a partner to run. We would co-develop it with them and other partners, but when it came time to lead, we would hand it over to someone else and say, 'We're going to be on the bus, but you're going to drive it.' If they're truly national missions, it shouldn't matter who leads them because it shouldn't change the outcome. But if your first step is taking ownership of the mission, then the mission is not likely to be successful.

Given the risk of failure in a bold, ambitious mission, the people who are volunteering to lead are the ones who want them to work. They've got to be collaborative missions with no individual owning them, but someone does need to do the monitoring and the management and be accountable for it. Managing missions is not about taking credit for their success; it's about being accountable for their failure. If you reframe who leads the mission as the one who's going to get blamed if it fails, then there's a far shorter line of people volunteering for that job. Just as we did when we proposed initiatives in the National Innovation and Science Agenda (NISA, discussed in Chapter 6), CSIRO will often offer to be accountable when initiatives don't work and support others to take credit when they do work, but that does mean CSIRO needs to be involved in the co-creation of missions to begin with.

Missions inspire us with common purpose

In Chapter 5, I discussed CSIRO's shift from referring to our 'market vision' to instead framing our 'six national challenges' to appeal to our purpose-driven people. I have always preferred to use the word 'startup' to refer to specific, goal-oriented ambitions – but that word had been about as popular in CSIRO as 'market vision'. But with the introduction of our missions program, I could reframe startups as 'missions', and the willingness of CSIRO people to take risks for missions has been a 180° swing, because they know this is really important. It's tapping into that 'save the world' mindset that drives scientists and engineers. They're willing to forget about the bureaucracy for a minute and focus on how we make this work. Quite quickly we saw scientists from different parts of CSIRO coming together to talk about ideas for missions in ways that would have otherwise been hard to catalyse. The missions program started to work because of the groundswell of support across Team CSIRO and the people who wanted to do it, not just because leaders told them to, which never really works at an organisation of smart, free-thinkers like at CSIRO.

Organisations face barriers created by their culture and history that they need to overcome; the system has barriers created by a very complex culture and a very complex history it needs to overcome. I think the only way to drive cultural change in the face of complexity, whether in a company or an innovation system, is by finding the motivation to change. Systems and processes will only get you so far. Let's solve a problem that's going to impact on our children and their children if we don't solve it. Missions create a 'why' we should change, not a 'how' we should change. The 'how' will always be messy in missions, because innovation is not linear, but if you know why you're doing it and you know what you want to do, that's a powerful catalyst for change. Only something that can drive widespread cultural change will be able to overcome the bear traps and buzzsaws left behind by the generations who built this dysfunctional system.

Embracing failure

Missions have a clear goal with no prizes for coming second, like the Hydrogen Industry Mission's goal of getting hydrogen to under $2 per kilogram by 2030. If they miss that target, there is no second prize because the market is simply not viable at any other price point. You can write all the research papers you like, but if you don't meet that goal and your product doesn't work, you fail like any other startup would.

The balancing act of developing a mission is seeing far enough into the future that you've got time to invent the science you need to solve a roadblock, but not so far that it's not fundable. We crowdsourced the brains trust of CSIRO and our partners for mission proposals and got an incredible number of ideas from people for things that we could do. When you are a VC partner hearing hundreds of pitches, you often wish you could assemble dream teams and dream products by cherry-picking from different startups – but you usually can't do it because of the egos involved. The beauty of the missions program was that we actually could do that within CSIRO. From the beginning idea to actual launch, there wasn't one mission that didn't merge teams and ideas, similar to the entrepreneurial behaviour we instilled in ON where we encouraged failure because it led to better outcomes.

Working on a mission guaranteed funding while you were sprinting towards launch, providing more security than a startup would have. But that also meant we had to be rigorous in cutting them off if they didn't deliver – exactly what would happen to a startup if it couldn't raise more money or start generating revenue. That made it wonderful when we had mission teams come to us and say, actually, we don't want to be missions anymore because we've failed.

One of those teams was working on a mission to double the number of Australian small and medium businesses using Australian science to develop their companies. It was a direct response to the industry side of Australia's innovation dilemma: the low interest of industry in funding deep-tech R&D. It was a huge, mission-like goal, but what they ultimately realised was that this goal should be at the heart of everything CSIRO does, not just in one specialised team. It should be an ambition that is driven out of every part of the organisation, so they took it back to their leadership team and used their insights to drive the entire organisation's small and medium business engagement strategy. This was an epic moment for me. I have always worried that setting KPIs will inadvertently drive the wrong behaviour, but here was a team making the right strategic decisions even though it hurt them and our board-imposed KPIs. This is one of the many reasons I love CSIRO people!

Another one of the missions in development was in health. The more the team talked to customers and understood the market, the more they realised their goal was too big for one mission. As the team had been keeping us updated on their progress, we could see they were deviating further and further from that ideal growth pattern that you see in startups (discussed in Chapter 2 and explained further for missions later in this chapter), but the team ended up coming to us as the leadership team before we could intervene. They acknowledged they'd wandered a bit too far from their original goal and couldn't see how to converge everything back into focus. But that attracted more of our business units to lean in to help and led to them using their collective insights and goals to completely redefine CSIRO's overarching health strategy. I had never seen anything like this in a corporation or government organisation. Twenty years earlier, former CSIRO Chief Executive Geoff Garrett had coined the phrase 'One CSIRO', and today we are finally seeing it in action.

The fast-fail sprints of the mission program have been important to work out what's actually a mission, but they've also been invaluable in the learnings and insights they've provided to the organisation as a whole. We might have got there later through our other planning and review processes, but never as quickly, and so much of a strategy's success is how fast it can seize a perishable opportunity. If they acted like the rest of CSIRO, mission teams would have had big ideas and they would have tried this, that or the next thing so that it would have just percolated under the radar for a couple of years before they got to the convergence point. But because they're on the missions program, we're monitoring them in a very particular way: in a higher-pressure environment with really fast cycles and not too much bureaucracy or control. That kind of integrity and the willingness to give up the opportunity to ride on the missions budget – what a great way to drive change in an organisation! Outside of a purpose-driven organisation, you'd be hard pressed to see that from your people.

Reimagining metrics and KPIs

Considering the challenges of how we measure innovation in Australia (discussed in Chapter 1), missions offer a fresh approach to tracking success without impinging on their need for agility. If you can't have business plans for fear of mutating the mission, minimising it or making it more risk-averse, how do you actually measure their results and manage them? If you're an internet company, there's a certain pattern to your revenue growth that people have analysed. I mentioned in Chapter 2 that Niki Scevak applied a similar approach to the investments he proposed when we created Blackbird. Many venture capitalists who invest in internet startups have a model for what the revenue should look like and they know when the best time to invest is on that revenue growth curve, so it's quite analytical. Believe it or not, missions have a similar fingerprint. Even though they are often working across completely different industries, markets, technologies and regions, they do have a pattern – but it's not as easy as measuring revenue. Steve Blank, founder of I-Corps, said, 'One of the key characteristics of good entrepreneurs is their ability to see "over the horizon" or "through the fog of war" – I just call it good pattern recognition skills.'[268]

In my mind there's a big difference between metrics and KPIs. KPIs have to be met; metrics have to be measured to guide you. Your metrics tell you if you are on course or not, allowing you to course-correct if not. There are about 10 metrics that we look at and monitor across a mission. If you just look at those metrics, they all look different and random, but if you add them together that's your total score – like going to the lolly shop and getting 5 cents' worth of one lolly and 10 cents' worth of another, then stacking them all up in one jar to reach a certain capacity. One mission may have a lot of green lollies, which are IP, filing patents; another mission may have none, but they have a lot of red lollies, which is customer engagement, so they're closer to a product vision. If you consolidate the gains across those 10 metrics, you get a curve that's almost the same for every mission. It's not something any of us expected, but having a fingerprint or a pattern for a mission is certainly something we'd hoped for. It is, of course, that entrepreneur's version of Metcalfe's law again – innovation isn't linear, collaboration is exponential. Missions harness the exponential power of networks and that's why they work so well.

That pattern is so important because you need to give your people enough agility to fail, learn, regroup and go in a different direction, multiple times. The path from idea to success is never a straight line, and the departure from that straight line gets bigger and bigger and bigger before it starts to get any smaller, so the more agile the mission is, the more radically it looks like it's out of control. If you give your people this same agility, you need to know when a mission actually is failing because it's unlikely they'll tell you if it is. If they're good entrepreneurs, they'll succeed or they'll die trying. They'll make the impossible possible to be successful, but they won't tell you that they're failing. I suspect the two CSIRO missions that left the missions program to be a broader part of CSIRO – one to support small business, one in health – only did so because they had found other non-missions pathways to success. I don't believe they would have abandoned their goals altogether.

Being able to measure a mission means you know when something isn't looking right or when something's missing. The measurements don't tell you what is missing exactly, but they'll draw your attention, encouraging you to go deeper with the team to try to figure it out, like you would with any other problem. Ultimately, you hope for consensus on the next steps. This is why missions must be co-created and why it's so powerful for Australia's missions to draw the bulk of their funding from industry instead of government, because from the start you have buy-in from which to continue seeking consensus.

Launching missions in a pandemic

We began preparing to launch our program of missions in 2020 – just as the global COVID-19 pandemic was beginning. Across CSIRO and many Australian universities, scientists dropped their research projects to join the global coronavirus effort. For our part, CSIRO researchers at the Australian Centre for Disease Preparedness in Geelong, Victoria, as well as at our sites around the country, turned to challenges as wide-ranging as studying the virus itself, developing and testing vaccine candidates, scaling up vaccine manufacturing capabilities, driving agile manufacturing that saw companies pivot to production lines of personal protective equipment, rolling out wastewater testing programs, analysing huge datasets of testing and case numbers, and addressing many other urgent questions to support Australia's response. For the first time in decades, science was on the front page of every newspaper and the headline on every news bulletin. It engaged people with the same hope and optimism that President Kennedy knew the Moon mission would inspire in the 1960s – albeit COVID-19 was a far darker and more tragic developing story. Although there was nothing mission-like about the planning or preparation for this unprecedented task, it was a true 'Team Australia' mission in the way silos dissolved, interorganisational relationships flourished, and information and knowledge were shared freely towards a mutual goal.

Recognising the unprecedented collaborative spirit that COVID-19 had unlocked across the innovation system, CSIRO launched its program of missions in August 2020 with an intention to form collaborative partnerships across government, industry, research and the community to drive Australia's economic recovery from the pandemic. We committed to investing at least $100 million annually,[269] with a goal of shifting one-third of CSIRO to be working on missions within the decade. It's important to note that, unlike the CRC program, this funding is just to cover CSIRO's involvement – not to incentivise any other participants. The merits of the mission had to stand on their own feet to attract partners.

I announced the program at the National Press Club in Canberra in August 2020:

> There are moments in time that shape generations – the generation who harnessed electricity; the generation who saw humans walk on the Moon; the generation who switched on the internet. Moments driven by science.
>
> This generation is living through a perfect storm of bushfires, pandemic and recession. Never in our lifetime has a country – or the world – turned to scientists in the way they are now. This is our moment. Our moment in time that will shape our generation. History will judge us by what we do next and our children will live with the consequences. Science has the unique and wonderful ability to unite people around a mission to achieve things that were once thought impossible.
>
> The 1969 Moon Mission galvanised the public, inspired rafts of other inventions we now use in our everyday lives and achieved something no one thought possible when it was announced. The rocket may have launched in the United States, but Australia enabled the mission through our Dish [Murriyang] in Parkes, which was listed on the Heritage Register this week and NASA tracking stations across Australia.

The 1988 Global Polio Eradication Initiative launched a global mission to eliminate polio and by 2016 reduced the number of people paralysed by 99.99 per cent. Again Australia was a leader in this global mission, becoming polio-free 16 years earlier in 2000, the same year Sydney hosted the Olympics.

Today we have that same opportunity to rally research, industry and community, around a new mission enabled by science – a mission of recovery and resilience ...

The last 30 years of economic growth have lulled us into a false sense of security and over time business investment in R&D has fallen. But there's nothing like a crisis to snap us back into action. Ironically, this recession may be the greatest opportunity for innovation-led growth we've had in decades.[270]

We were joined in our announcement that day with support for the missions from then-Minister for Industry, Science and Technology Karen Andrews, Australia's then-Chief Scientist Dr Alan Finkel, Australian National University Vice-Chancellor Professor Brian Schmidt and v2food founder and then-CEO Nick Hazell[271] – just a sample of the partners we were engaging across the 12 missions in development.

Each of these missions in development was actively building partnerships and honing technological timeframes and goals to tackle specific market-based obstacles. At the time we didn't announce their time-bound, specific goal statements, because it was important that they be co-developed, so this is the list we published:[272]

- a *Pandemic Mission* to increase our resilience and preparedness against infectious diseases and pandemics

- a *Disaster Mission* to mitigate the impact of disasters: drought, bushfires and floods

- a *Hydrogen Mission* to create a hydrogen industry to generate a new clean energy export industry

- a *Manufacturing Mission* to accelerate the transition to agile manufacturing for higher revenue and sovereign supply

- an *Antimicrobial Resistance Mission* to overcome our growing resistance to antibiotics, so they keep saving lives

- a *Navigating Climate Change Mission* to create a national climate capability to navigate climate change uncertainty

- a *Drought Mission* to help our farmers overcome drought, mitigate climate impacts, and increase yield and profitability

- a *Future Protein Mission* to create a sustainable future protein industry

- a *Trusted AgriFood Exports Mission* to leverage the world's love of Australian-grown food to collectively drive our trusted agriculture and food exports to $100 billion
- a *Net Zero Mission* to use technology to navigate Australia's transition to net zero emissions, without derailing our economy
- an *AquaWatch Mission* to safeguard the health of our waterways by monitoring the quality of our water resources from space
- a *Critical Energy Metals Mission* to create new industries that transform raw mineral commodities into unique higher-value products like critical energy metals that build Australia's value-added offering, jobs and sovereign supply
- an *Ending Plastic Waste Mission* to reinvent the way plastic is made, processed and recycled
- an *SME Collaboration Nation Mission* to double the number of SMEs benefiting from Australian science.

Over the following 2 years, we launched six of these missions, completely co-developed with a broad coalition of partners and announced at the time as follows:

- a Hydrogen Industry Mission to scale demand and drive down hydrogen cost to under $2 per kilogram by 2030.[273] More than 100 projects worth $68 million have been planned by partners including the Department of Industry, Science, Energy and Resources (DISER), Australian Renewable Energy Agency (ARENA), Fortescue Metals Group, Swinburne University, the Victorian Government, the Future Fuels CRC, National Energy Resources Australia (NERA), and the Australian Hydrogen Council, along with collaborators Toyota and Hyundai. CSIRO and Boeing – research partners for more than 30 years – will also continue to explore hydrogen's future use in the aviation industry.
- a Drought Resilience Mission to reduce the impacts of drought by 30 per cent by 2030.[274] It brings together CSIRO, Department of Agriculture, Water and the Environment and Bureau of Meteorology.
- a Trusted Agrifood Exports Mission to increase the value of Australian agrifood exports by $10 billion by 2030.[275] It has live projects underway with the Department of Agriculture, Water and Environment, Hort Innovation and Meat and Livestock Australia.
- a Future Protein Mission to produce an additional $10 billion of high-quality protein products by 2030.[276] It brings together CSIRO with

Department of Industry, Science, Energy and Resources, Victorian State Government, Meat & Livestock Australia and the Grains Research & Development Corporation along with industry partners v2food, GrainCorp, Ridley, Clara Foods, Wide Open Agriculture and startups such as Eden Brew. This mission is also working with Food Innovation Australia Limited; Austrade; Victorian, New South Wales and South Australian state governments; Queensland's Department of Agriculture and Fisheries; Western Australia's Department of Primary Industries and Regional Development; and the University of Queensland, University of Sydney, University of Technology Sydney, University of NSW and Edith Cowan University.

- an Ending Plastic Waste Mission to reduce Australia's plastic waste by 80 per cent by 2030.[277] It brings together CSIRO with government partners Department of Agriculture, Water and the Environment; Department of Foreign Affairs & Trade; Department of Industry, Science, Energy and Resources; NSW Department of Planning Industry and Environment; Parks Australia; NSW Environmental Trust and Sustainability Victoria; along with industry partners Australian Packaging Covenant Organisation, Chemistry Australia, C Sea Solutions, Kimberly-Clark Australia, Circle 8 Clean Technologies, Ostrom Polymers and Phantm; and a number of universities including Monash University, Murdoch University, University of Technology Sydney, University of Queensland, RMIT University, University of Tasmania, and University of Western Australia. The mission is also partnering with a number of international organisations including the NSF, University of Texas at Austin and the US Department of Energy's BOTTLE Consortium.

- a Towards Net Zero Mission to reduce the emissions of some of Australia's hardest to abate industries by 50 per cent by 2035.[278] It brings together CSIRO with government partners and collaborators Climate Change Authority; Department of Climate Change, Energy; Environment and Water; Department of Industry, Science and Resources; Queensland Department of Agriculture and Fisheries; along with industry partners and collaborators BHP; Boeing; Climate Leaders Coalition (CLC); Climate-Kic; ClimateWorks; Incitec Pivot; KPMG Australia; Meat & Livestock Australia; Qantas; and a number of universities and research organisations including Heavy Industry Low-carbon Transition Cooperative Research Centre.

I'm confident the missions program has a bright future, both inside and outside of CSIRO. As Emeritus Professor Roy Green, Special Innovation Advisor at the University of Technology Sydney, says, the mission model ultimately needs to be a national program: 'The idea of the missions is great and they should be a national endeavour right across research and innovation, not just in CSIRO, compartmentalising it in CSIRO isn't adequate. It was a good experiment to do it in there, but it now needs to go much further.'[279]

Australia's future missions

Missions designed for Australia can transform our innovation system to be more aligned, focused and collaborative. They can bring partners across industry, research, government and the community together around bold, ambitious, time-bound and specific goals without the need for 'attaching strings' to funding for research. Unlike earlier research–industry coordination efforts, they also invite the venture community to drive the creation of startups. Missions have clear, time-bound goals, the same way startups do; this will appeal to the venture community more broadly as they begin to see the pipeline of startups being formed by missions.

I'm excited by the missions model that CSIRO is growing and believe these missions can drive transformational, system-wide change. Former CSIRO Chair David Thodey said the organisation is unique in its ability to drive missions because its 'collective and multidisciplinary capabilities are unique and differentiate CSIRO from many other organisations. CSIRO is a unique and special "gem" in Australia and is a world-class and renowned global research organisation.'[280] With economic conditions tipped to continue to tighten, industry may have other short-term investment priorities although, as discussed in Chapter 2, deep-tech doesn't operate on the same cycles as the rest of the market.

All the interventions discussed in this book come together in a uniquely Australian model for missions, which have the most potential for significant impact because of the depth and breadth of the partners they bring together. At the individual level, there are incentives for venture capital to invest in startups led by scientist CEOs with a market vision that grows the pie for everyone. At a company level, they align organisations around a common set of national challenges and tackle deep-tech innovation reticence in corporate Australia through the opportunity for venture science. All of these flow through across the innovation system, creating a unique and vibrant Australian environment where our science excellence is translated into real-world solutions that grow a prosperous and sustainable future.

Conclusion: leaping forward

As I complete my time leading our national science agency, CSIRO, and reflect on my three decades of deep-tech innovation, I have never been more optimistic about our country's innovation future. Yes, we have barriers to innovation, including social barriers like complacency from our mineral wealth, cultural cringes around careers and competition in a 'finite pie' mentality, as well as systemic barriers like not always measuring or incentivising the right things, not backing a market vision for our science and the small size, youth and risk aversion of our investment and venture system. But we have done amazing things with few resources, so I am excited about what's possible as we start to see more VC flowing and more deep-tech founders and CEOs finding their feet.

It does a disservice to our great Australian culture to suggest we should copy Silicon Valley or any other innovation system around the world – we just need to make it that little bit easier to leap across the Valley of Death. Not everyone will make it across – and nor should they. The market operates the way it does for a reason. If anything, we should expect to see more failed startups strewn across that Valley – because it's only by supporting more deep-tech startups to make that leap that more will land successfully on the other side and flourish, luring more and more to follow their path.

I see Australia's digital companies thriving and I believe our deep-tech startups can – and must – follow swiftly in their footsteps, so they can invent the next generation of breakthrough technologies that, in turn, drive the next generation of digital companies, in a complementary dance. I see our venture system growing in size and risk appetite and I believe they will be richly rewarded for investing in our brightest minds tackling our greatest challenges. And I see our purpose-driven scientists leaning into their latent entrepreneurial skills to push their inventions out into the world and I believe they will be inspiring and powerful role models to the next generation.

My market vision for Australia builds on these important foundations to see a uniquely Australian innovation system that doesn't relegate our biggest companies to the role of 'commercialiser', but instead elevates them to partner, investor and 'cornerstone', through models like venture science and in coalitions formed around shared goals like our missions. Our businesses of all sizes – from corporates to startups – will be our weathervanes for success as we grow jobs and economic value, as well as our guiding star for investment as they show us where Australia can have an unfair advantage and compete globally.

The stage is already set for us to write the next chapter of our innovation story. It will have the same brilliant Australian cast, but, instead of being an expat's story told from the US, China or Israel, it's a homegrown champion's story. Instead of wi-fi going to the US, Morse Micro grows here; instead of SunTech going to China, Solar Drive grows here; instead of Better Place Australia being dragged down by global headquarters, Tritium grows here. After receiving the first images from the Moon via NASA, we send satellites up with Gilmour Space. After digging up our raw materials and exporting them to the world, we power remote communities with hydrogen generation and storage solutions with Endua and export them to the world. After feeding so much of the world with our beef, we grow the global protein market with v2food and reduce the environmental footprint with FutureFeed.

We already have the skills, tools and resources we need to create a new, sustainable and rewarding economic pillar from our science: the purpose-led scientists with brilliant ideas, the visionary investors with access to a healthy capital market, and the growing pool of scientist CEOs who are redefining what leadership looks like in the innovation age. Working together, we can nudge each other out of our comfort zones of invention and into the opportunities of innovation. It's more than a leap of faith – it's a leap of science.

Acknowledgements

DR LARRY MARSHALL

The idea for this book was born when I realised I had more to say about innovation than could ever fit in a single speech. I passionately believe we have some of the world's best scientists in Australia and that they hold the keys to solving our greatest challenges and unlocking a better future for all of us. I dedicate this book to them.

I am indebted to the many experts who gave their time and wisdom through interviews, emails and conversations for this book, both reflecting on our shared experiences and contributing new ideas and insights. We were only able to include a small fraction of their collective genius, but their actions and ongoing contributions to strengthening Australia's innovation system speak louder than any book ever could.

Thank you to my PhD advisor Professor Jim Piper and examiner Professor Bob Byer. Thank you to every member of my six startups, without your passionate persistence we would have failed. Thank you to my team at startup number seven, Team CSIRO, for teaching and inspiring me every day, especially my amazing and award-winning office team for their tireless support, good humour and incisive judgement: Annemaree Lonergan, Beth Cribb, Jack Steele, Mark Bazzacco, Philippa Eggerton, Sascha Burmester and Tracey Cootes.

Finally and most importantly, thank you Maria, Patrick and Jessica for tolerating three decades of travel, six startups and two VC Funds, CSIRO and endless politics and media, and my mission to solve Australia's innovation dilemma – I don't know how you put up with it all but I am so lucky you did and your support means the world to me.

JENNA DAROCZY

Thank you, Larry, for accepting that one speech cannot, in fact, contain all your brilliant ideas, and for inviting me on this wonderful book-writing adventure! Thank you, Tanya Bowes, Crystal Ladiges and the CSIRO Publishing team, for being so supportive of this experiment. And to my beloved friends and family, thank you for cheering me on, waiting patiently for me to emerge from hibernation and knowing how much I've always wanted to do this.

The authors acknowledge the Traditional Owners of the lands across Australia and pay their respect to Elders past and present. They recognise that Aboriginal and Torres Strait Islander peoples have made and will continue to make extraordinary contributions to all aspects of Australian life including culture, economy and science.

Abbreviations

AI: artificial intelligence

ASX: Australian Securities Exchange

CHIPS and Science Act: Creating Helpful Incentives to Produce Semiconductors and Science Act

CRCs: Cooperative Research Centres

DARPA: Defense Advanced Research Projects Agency

DAS: Digital Agriculture Services

DFJ: Draper Fisher Jurvetson

ESVCLP: early stage venture capital limited partnership

GFC: Global Financial Crisis

IIF: Innovation Investment Fund

IP: intellectual property

IPO: initial public offering

KPI: key performance indicator

MBA: Master of Business Administration

MRD: Market Requirements Document

NASDAQ: National Association of Securities Dealers Automated Quotations

NCI: normalised citation impact

NICTA: National Information and Communications Technology Australia

NISA: National Innovation and Science Agenda

NSF: National Science Foundation

OECD: Organisation for Economic Co-operation and Development

PRD: Product Requirements Document

R&D: research and development

RDTI: Research and Development Tax Incentive

RIO: Redfern Integrated Optics

SBIR: Small Business Innovation Research

SMEs: small and medium enterprises

STEM: science, technology, engineering and mathematics

TRL: technology readiness level

VC: venture capital

WIPO: World Intellectual Property Organization

Endnotes

1   https://csiropedia.csiro.au/wlan-the-next-generation-1993/
2   https://data61.csiro.au/en/Our-Research/Our-Work/Future-Cities/Planning-
     sustainable-infrastructure/WiFi
3   https://www.ipaustralia.gov.au/tools-resources/case-studies/csiro-wlan-patent
4   Email from Dr Jack Steele, 7 September 2022
5   https://www.mq.edu.au/research/research-expertise/Research-innovation/
     where-wi-fi-began
6   Email from Dave Skellern, 17 November 2022
7   Email from Dave Skellern, 17 November 2022
8   Interview with Dave Skellern, 22 October 2022
9   https://www.mq.edu.au/research/research-expertise/Research-innovation/
     where-wi-fi-began
10   https://www.ipaustralia.gov.au/tools-resources/case-studies/csiro-wlan-patent
11   https://sief.org.au/wp-content/uploads/2019/01/SIEF-Annual-Report-2009-10.pdf
12   https://www.csiro.au/en/news/News-releases/2015/CSIRO-fund-to-support-
     Australian-startups
13   https://www.nature.com/nature-index/annual-tables/2022/country/all/all
14   https://csiropedia.csiro.au/prickly-pear-control/
15   https://blog.csiro.au/larry-marshalls-national-press-club-address/
16   Interview with Catherine Livingstone, 18 August 2022
17   Interview with Steven Worrall, 30 August 2022
18   https://publications.industry.gov.au/publications/
     australianinnovationsystemmonitor/index.html
19   https://publications.industry.gov.au/publications/
     australianinnovationsystemmonitor/science-and-research/business-R-and-D/
     index.html
20   CSIRO (2022) CSIRO Strategy Corporate Plan 2022–23, p. 18,
     https://www.csiro.au/-/media/About/Corporate-Plan/Corporate-
     Plan-22-23/22-00137_CORP_REPORT_CorporatePlan2022-23_WEB_220830.pdf
21   CSIRO (2022) CSIRO Strategy Corporate Plan 2022–23, p. 17,
     https://www.csiro.au/-/media/About/Corporate-Plan/Corporate-
     Plan-22-23/22-00137_CORP_REPORT_CorporatePlan2022-23_WEB_220830.pdf
22   https://publications.industry.gov.au/publications/
     australianinnovationsystemmonitor/science-and-research/business-R-and-D/
     index.html

23 AlphaBeta (2018) 'Digital innovation: Australia's $315b opportunity', https://data61.csiro.au/~/media/D61/Files/Digital-Innovation-Report.pdf

24 Department of Industry, Science, Energy and Resources (2020) Australian Innovation System Monitor, https://www.dese.gov.au/about-us/resources/university-research-commercialisation-consultation-paper

25 Organisation for Economic Co-operation and Development, Main Science and Technology Indicators, Vol. 2019/2, https://www.dese.gov.au/about-us/resources/university-research-commercialisation-consultation-paper

26 https://www.wipo.int/edocs/pubdocs/en/wipo_gii_2015.pdf

27 Australia's Chief Scientist (2015) 'STEM skills in the workforce: what do employers want?', https://www.chiefscientist.gov.au/2015/04/occasional-paper-stem-skills-in-the-workforce-what-do-employers-want

28 Consult Australia (2019) Australia's STEM Education Challenges, https://www.consultaustralia.com.au/docs/default-source/people/people-page/australia's-stem-education-challenges-discussion-paper.pdf?sfvrsn=652a4ab9_2

29 Organisation for Economic Co-operation and Development (2019) Programme for International Student Assessment (PISA), 2018

30 Consult Australia (2019) Australia's STEM Education Challenges, https://www.consultaustralia.com.au/docs/default-source/people/people-page/australia's-stem-education-challenges-discussion-paper.pdf?sfvrsn=652a4ab9_2

31 https://publications.industry.gov.au/publications/australianinnovationsystemmonitor/science-and-research/research-output/index.html

32 Knowledge Commercialisation Australasia (2020) Survey of Commercialisation Outcomes from Public Research (SCOPR), https://www.dese.gov.au/about-us/resources/university-research-commercialisation-consultation-paper

33 Knowledge Commercialisation Australia (KCA) Survey of Commercialisation Outcomes from Public Research (SCOPR) 2021 Annual Report

34 Stanford total revenue ($12.5bn) from Stanford's Annual Report: https://bondholder-information.stanford.edu/sites/g/files/sbiybj21416/files/media/file/stanford-annual-financial-report-2020.pdf
Stanford licensing revenue ($114m) from: https://facts.stanford.edu/research/innovation/

35 Interview with Roy Green, 23 August 2022

36 Interview with Catherine Livingstone, 18 August 2022

37 Email from Andrew Stevens, 5 September 2022

38 https://www.ipaustralia.gov.au/sites/default/files/ip_report-2022-may.pdf, p. 10

39 https://www.baraja.com/en/company

40   https://www.afr.com/young-rich/it-started-in-a-tiny-garage-now-this-car-tech-company-is-worth-300m-20210913-p58r84

41   Email from Phil Morle, 6 September 2022

42   Interview with Andrew Forrest, 12 August 2022

43   Turnbull M (2020) *A Bigger Picture*, Hardie Grant Books, Richmond, p. 306

44   Interview with Matt Barrie, 19 July 2022

45   Interview with Josh Makower, 27 July 2022

46   Interview with Josh Makower, 27 July 2022

47   Interview with Steve Blank, 29 July 2022

48   Interview with Niki Scevak, 15 September 2022

49   Email from Genevieve Bell, 23 August 2022

50   Interview with Andrew Forrest, 12 August 2022

51   https://www.csiro.au/en/news/News-releases/2022/Steering-Australia-through-the-Valley-of-Death

52   https://www.oecd-ilibrary.org/sites/2bf6bc72-en/index.html?itemId=/content/component/2bf6bc72-en#

53   Interview with Catherine Livingstone, 18 August 2022

54   Interview with Raphael Arndt, 16 August 2022

55   Interview with Alan Finkel and Evan Thornley, 12 August 2022

56   https://www.rba.gov.au/publications/submissions/financial-sector/background-on-the-australian-listed-equity-market-2021-09/index.html

57   Interview with Niki Scevak, 15 September 2022

58   Interview with Tristen Langley, 19 July 2022

59   Marshall L (16 August 2017) 'Incubate a little culture change', *The Australian*, https://www.theaustralian.com.au/higher-education/incubate-a-little-culture-change/news-story/4431544ec5025497553eb85ca766dd41

60   Interview with Simon McKeon, 9 August 2022

61   Interview with Tristen Langley, 19 July 2022

62   Interview with Raphael Arndt, 16 August 2022

63   Interview with Matt Barrie, 19 July 2022

64   Interview with Dean McEvoy, 16 August 2022

65   Interview with Steve Blank, 29 July 2022

66   Email from Robyn Denholm, 24 August 2022

67   Interview with David Thodey, 8 August 2022

68   Interview with Simon McKeon, 9 August 2022

69   Interview with Roy Green, 23 August 2022

70   Email from Genevieve Bell, 23 August 2022

71   Email from Phil Morle, 6 September 2022

72   https://www.nature.com/nature-index/annual-tables/2022/country/all/all

73   Interview with Roy Green, 23 August 2022

74   Interview with Steven Worrall, 30 August 2022

75   https://treasury.gov.au/publication/economic-roundup-issue-2-2011/economic-roundup-issue-2-2011/the-resources-boom-and-structural-change-in-the-australian-economy

76   https://www.csiro.au/en/news/News-releases/2022/Speech-to-the-National-Press-Club---Launch-of-the-2022-Our-Future-World-report

77   Department of Industry, Science, Energy and Resources (2017) Australian Innovation System Report, 2017, https://www.industry.gov.au/data-and-publications/australian-innovation-system-report/australian-innovation-system-report-2017

78   Email from Atlassian, 13 September 2022

79   Interview with Niki Scevak, 15 September 2022

80   Email from Atlassian, 13 September 2022

81   Interview with Dean McEvoy, 16 August 2022

82   Interview with Matt Barrie, 19 July 2022

83   Interview with Catherine Livingstone, 18 August 2022

84   Ferris B (1999) *Nothing Ventured, Nothing Gained: Thrills and Spills in Venture Capital*, Allen & Unwin, Sydney, p. 94

85   Ferris B (1999) *Nothing Ventured, Nothing Gained: Thrills and Spills in Venture Capital*, Allen & Unwin, Sydney, p. 94

86   *Extraordinary Measures* (2010), Tom Vaughan, Sony Pictures Releasing International

87   Ferris B (1999) *Nothing Ventured, Nothing Gained: Thrills and Spills in Venture Capital*, Allen & Unwin, Sydney, p. 74

88   Interview with Catherine Livingstone, 18 August 2022

89   Interview with Niki Scevak, 15 September 2022

90   Interview with Tristen Langley, 19 July 2022

91   Interview with Matt Barrie, 19 July 2022

92   Interview with Raphael Arndt, 16 August 2022

93   https://www.aic.co/common/Uploaded%20files/Case%20Studies/Jodie%20Fox_Shoes%20of%20Prey.pdf

94   Interview with Raphael Arndt, 16 August 2022

95   Interview with Raphael Arndt, 16 August 2022

96   Interview with Niki Scevak, 15 September 2022

97   Interview with Niki Scevak, 15 September 2022

98   https://www.forbes.com/sites/tomtaulli/2020/06/27/amazon-buys-zoox-why-self-driving-technology-is-existential/?sh=64909c8b7c0f

99   Email from Niki Scevak on 10 October 2022, with further reference to https://www.blackbird.vc/investors

100 Interview with Niki Scevak, 15 September 2022

101 Interview with Matt Barrie 19 July 2022

102 Interview with Evan Thornley and Alan Finkel, 12 August 2022

103 Interview with Niki Scevak, 15 September 2022

104 Interview with Simon McKeon, 9 August 2022

105 Interview with Raphael Arndt, 16 August 2022

106 Interview with Sam Sicilia, 22 August 2022

107 https://guides.loc.gov/wall-street-history/exchanges

108 Ferris B (1999) *Nothing Ventured, Nothing Gained: Thrills and Spills in Venture Capital*, Allen & Unwin, Sydney, pp. 100–101

109 Interview with Matt Barrie, 19 July 2022

110 Interview with Richard Favero, 24 August 2022

111 Interview with Evan Thornley and Alan Finkel, 12 August 2022

112 Interview with Dean McEvoy, 16 August 2022

113 https://hbr.org/2019/11/the-truth-about-ceo-tenure

114 https://www.theage.com.au/national/victoria/race-to-be-qantas-boss-takes-off-20220309-p5a390.html

115 Interview with Niki Scevak, 15 September 2022

116 Interview with Tristen Langley, 19 July 2022

117 Interview with Simon McKeon, 9 August 2022

118 Interview with Raphael Arndt, 16 August

119 Email from Robyn Denholm, 24 August 2022

120 Robyn Denholm speaking at the AFR Business Summit in March 2022, https://www.afr.com/business-summit/australia-s-next-unicorns-will-come-from-five-areas-20220311-p5a3tl

121 Interview with Steven Worrall, 30 August 2022

122 https://www.investopedia.com/articles/investing/102615/story-instagram-rise-1-photo0sharing-app.asp

123 http://www.sxvp.com/fund-1

124 https://www.nokia.com/about-us/news/releases/2014/06/05/nokia-buys-compact-radio-systems-expert-mesaplexx/

125 https://www.aic.co/common/Uploaded%20files/Case%20Studies/Jodie%20Fox_Shoes%20of%20Prey.pdf

126 Wozniak S, Smith G (2006) *iWoz: The Autobiography of the Man Who Started the Computer Revolution*, Headline Review, UK, pp. 157–158

127 Interview with Dean McEvoy, 16 August 2022

128 Interview with David Thodey, 8 August 2022

129 Interview with Andrew Forrest, 12 August 2022

130 Email correspondence with Chauncey Shey, November 2022

131 Wozniak S, Smith G (2006) *iWoz: The Autobiography of the Man Who Started the Computer Revolution*, Headline Review, UK, pp. 239–241

132 Email correspondence with Bill Ferris, October 2022

133 https://www.smh.com.au/technology/tax-policy-hitting-entrepreneurs-20120827-24wex.html

134 https://www.mercurynews.com/2014/07/17/wife-of-google-co-founder-acknowledges-role-in-downtown-los-altos-remake/

135 Email from Atlassian, 13 September 2022

136 https://www.afr.com/wealth/people/canva-s-billion-dollar-giveaway-will-be-nation-s-most-generous-20210915-p58rvj

137 https://ecos.csiro.au/global-sustainability-goals-a-challenge-for-australia/ and https://www.nature.com/articles/nature21694.epdf

138 https://www.bsu.edu/about/administrativeoffices/careercenter/tools-resources/personality-types/entp

139 https://www.bsu.edu/about/administrativeoffices/careercenter/tools-resources/personality-types/entj

140 https://www.afr.com/work-and-careers/careers/david-thodey-peter-botten-make-top-100-global-ceo-list-20151109-gkuqjr

141 Interview with David Thodey, 8 August 2022

142 Interview with David Thodey, 8 August 2022

143 https://en-academic.com/dic.nsf/enwiki/1699336

144 Email from Frank Levinson, 29 August 2022

145 Interview with Tristen Langley, 19 July 2022

146 Interview with Matt Barrie, 19 July 2022

147 Interview with Peter Farrell, 24 August 2022

148 Interview with Peter Farrell, 24 August 2022

149 Interview with Richard Favero, 24 August 2022

150 Interview with Richard Favero, 24 August 2022

151 Interview with Steve Blank, 29 July 2022

152 Interview with Niki Scevak, 15 September 2022

153 Thiel P, Masters B (2014) *Zero to One: Notes on Startups, or How to Build the Future*, Virgin Books, UK, p. 1

154 Email from Atlassian, 13 September 2022

155 Interview with Josh Makower, 27 July 2022

156 Email from Frank Levinson, 29 August 2022

157 Interview with Andrew Forrest, 12 July 2022

158 Interview with Andrew Forrest, 12 August 2022

159 Interview with Andrew Forrest, 12 August 2022

160 Interview with Dean McEvoy, 16 August 2022

161 https://www.deanmcevoy.com/

162 https://www.smartcompany.com.au/entrepreneurs/spreets-co-founder-dean-mcevoy-reveals-how-he-got-through-his-darkest-days-to-build-a-tech-company-that-sold-for-40-million/

163 Interview with Dean McEvoy, 16 August 2022

164 Marshall L (May 2001) Leaping into the void, *Laser Focus World*, https://www.optoelectronics-world.com

165 Wozniak S, Smith G (2006) *iWoz: The Autobiography of the Man Who Started the Computer Revolution*, Headline Review, UK, p. 251

166 The Osborne effect: https://www.pcmag.com/encyclopedia/term/osborne-effect

167 Interview with David Thodey, 8 August 2022

168 Interview with Andrew Forrest, 12 August 2022

169 https://www.reuters.com/article/infineracorp-ipo-idUSWEN863320070606

170 https://www.afr.com/technology/michelle-simmons-picked-up-46-million-to-build-a-computer-which-is-millions-of-times-quicker-20151222-glt8z2

171 Interview with Simon McKeon, 9 August 2022

172 Interview with Niki Scevak, 15 September 2022

173 Interview with Tristen Langley, 19 July 2022

174 Interview with Dean McEvoy, 16 August 2022

175 Interview with Alan Finkel and Evan Thornley, 12 August 2022

176 Interview with Alan Finkel and Evan Thornley, 12 August 2022

177 https://www.bloomberg.com/news/articles/2022-09-09/rivian-stock-price-rivn-jumps-on-mercedes-partnership?leadSource=uverify%20wall

178 Interview with Andrew Forrest, 12 August 2022

179 Interview with Josh Makower, 27 July 2022

180 Interview with Josh Makower, 27 July 2022

181 Interview with Peter Farrell, 24 August 2022

182 https://www.bizjournals.com/bizwomen/news/latest-news/2020/07/willow-breast-pump-startup-new-ceo-laura-chambers.html?page=all

183 Interview with Josh Makower, 27 July 2022

184 Interview with Josh Makower, 27 July 2022

185 Interview with Tristen Langley, 19 July 2022

186 https://medicalrepublic.com.au/telstra-buys-medical-director-it-all-changes-from-here/50566

187 Interview with Mary Foley, 29 August 2022

188 https://www.theaustralian.com.au/business/companies/digital-connectivity-key-as-mary-foley-dials-in-to-lead-telstras-health-kick/news-story/9f9336a4a37c6c2c43e7e76d02e10420

189 Email correspondence with Mary Foley, November 2022

190 Comments made by Minister Ed Husic during a panel conversation at the University of Technology, Sydney, Vice-Chancellor's Innovation Showcase event, 3 November 2022

191 https://www.csiro.au/en/news/news-releases/2018/csiros-cyanide-free-gold-showcases-non-toxic-solution

192 https://chrysos.com.au/

193 Interview with Andrew Forrest, 12 August 2022

194 Interview with Andrew Forrest, 12 August 2022

195 Email from Robyn Denholm, 24 August 2022

196 Interview with Steve Blank, 29 July 2022

197 https://www.afr.com/technology/meet-jack-ma-the-man-behind-alibaba-20140908-jeqeh

198 https://arena.gov.au/knowledge-bank/startup-spotlight-southern-cross-renewable-energy-fund/

199 https://newsroom.unsw.edu.au/news/science-tech/pioneering-solar-energy

200 https://www.eastpennmanufacturing.com/loving-memory-sally-s-miksiewicz/

201 Interview with Evan Thornley and Alan Finkel, 12 August 2022

202 https://www.csiro.au/~/media/About/Files/Strategy/CSIRO_Strategy_2020-PDF.pdf

203 Turnbull M (2020) *A Bigger Picture*, Hardie Grant Books, Richmond, pp. 279–280

204 https://pmtranscripts.pmc.gov.au/release/transcript-40115

205 Turnbull M (2020) *A Bigger Picture*, Hardie Grant Books, Richmond, pp. 151, 279–280

206 Interview with Steve Blank, 29 July 2022

207 Interview with Andrew Forrest, 12 August 2022

208 Email from Phil Morle, 6 September 2022

209 Interview with Steve Blank, 29 July 2022

210 Interview with Steve Blank, 29 July 2022

211 https://www.afr.com/technology/hostplus-temasek-lockheed-martin-back-csiros-main-sequence-ventures-20180911-h158vs

212 Interview with Sam Sicilia, 22 August 2022

213 Interview with Sam Sicilia, 22 August 2022

214 https://hostplus.com.au/news/hostplus-backs-main-sequence-second-fund

215 https://www.unisuper.com.au/about-us/media-centre/2022/unisuper-invests-with-uniseed

216 https://www.csiro.au/en/news/News-releases/2022/CSIROs-role-in-Australias-Economic-Accelerator-initiative

217 Interview with Roy Green, 23 August 2022

218 Provided by Dr Jack Steele, CSIRO, 11 and 17 August 2022

219 Interview with Raphael Arndt, 16 August 2022

220 Turnbull M (2020) *A Bigger Picture*, Hardie Grant Books, Richmond, p. 657

221 Turnbull M (2020) *A Bigger Picture*, Hardie Grant Books, Richmond, pp. 279–280

222 Interview with Catherine Livingstone, 18 August 2022

223 https://www.csiro.au/en/news/News-releases/2020/Adapt-to-megatrends-for-economic-recovery

224 https://www.abc.net.au/news/2020-08-08/coronavirus-boosts-telehealth-startup-coviu-business/12519234

225 https://www.investment.nsw.gov.au/living-working-and-business/nsw-innovation-and-productivity-council/our-publications/role-of-anchors/

226 https://www.csiro.au/en/work-with-us/services/consultancy-strategic-advice-services/csiro-futures/innovation-business-growth/australian-national-outlook

227 Interview with Roy Green, 23 August 2022

228 Interview with Peter Farrell, 24 August 2022

229 Interview with Steven Worrall, 30 August 2022

230 https://www.csiro.au/en/news/News-releases/2021/Google-Australia-announces-1-billion-Digital-Future-Initiative

231 https://www.csiro.au/en/news/News-releases/2021/CSIRO-and-Google-announce-five-year-partnership

232 Australian Academy of Technology and Engineering (ATSE) (2013) 2013 ATSE Clunies Ross Award: Dr Simon Poole and Dr Steven Frisken – Optical Communications, https://www.youtube.com/watch?v=AHQarwrXoy0

233 Email from Frank Levinson, 29 August 2022

234 Interview with Evan Thornley and Alan Finkel, 12 August 2022

235 https://newsroom.unsw.edu.au/news/science-tech/telstra-matches-10m-cba-pledge-quantum-computing-race

236 Augustin MA (2021) 'Perspective article: towards a new venture science model for transforming food systems', *Global Food Security* **28**, 100481, https://www.sciencedirect.com/science/article/pii/S2211912420301346?via%3Dihub

237 Augustin MA (2021) 'Perspective article: towards a new venture science model for transforming food systems', *Global Food Security* **28**, 100481, https://www.sciencedirect.com/science/article/pii/S2211912420301346?via%3Dihub

238 Interview with Catherine Livingstone, 18 August 2022

239 https://www.csiro.au/en/work-with-us/services/consultancy-strategic-advice-services/csiro-futures/futures-reports/agriculture-and-food/australias-protein-roadmap

240 Interview with Jack Cowin, 22 August 2022

241 Interview with Jack Cowin, 22 August 2022

242 Interview with Jack Cowin, 22 August 2022

243 https://www.future-feed.com/

244 https://www.mseq.vc/msv-challenges/decarbonise-the-planet

245 https://www.csiro.au/en/news/news-releases/2021/endua-to-build-next-gen-of-clean-hydrogen-energy-storage

246 https://www.csiro.au/en/news/news-releases/2021/endua-to-build-next-gen-of-clean-hydrogen-energy-storage

247 https://www.endua.com/team

248 https://www.csiro.au/en/news/news-releases/2021/space-startup-quasar-takes-off-with-csiro-tech

249 https://www.quasarsat.com/our-team

250 https://www.csiro.au/en/about/Corporate-governance/Ensuring-our-impact/Auditing-our-impact

251 https://csiropedia.csiro.au/radar/

252 https://www.dese.gov.au/about-us/resources/university-research-commercialisation-consultation-paper

253 Mazzucato M (2013) *The Entrepreneurial State: Debunking Public vs Private Sector Myths*, Anthem Press, London, p. 6

254 https://ec.europa.eu/research-and-innovation/en/horizon-magazine/missions-will-require-revolution-european-governments-prof-mariana-mazzucato

255 https://ec.europa.eu/info/news/commission-launches-work-major-research-and-innovation-missions-cancer-climate-oceans-and-soil-2019-jul-04_en

256 https://www.nature.com/articles/d41586-021-00496-z

257 Mazzucato M (2013) *The Entrepreneurial State: Debunking Public vs Private Sector Myths*, Anthem Press, London, p. 204

258 Interview with Roy Green, 23 August 2022

259 Email from Phil Morle, 6 September 2022

260 https://www.csiro.au/en/news/News-releases/2022/CSIRO-on-a-mission-to-end-plastic-waste

261 Interview with Steven Worrall, 30 August 2022

262 https://www.csiro.au/en/work-with-us/services/consultancy-strategic-advice-services/csiro-futures/innovation-business-growth/australian-national-outlook

263 Interview with Roy Green, 23 August 2022

264 Interview with Richard Favero, 24 August 2022

265 Interview with Roy Green, 23 August 2022

266 CSIRO Annual Report 2022, p. 43

267 https://digitalagricultureservices.com/company

268 Interview with Steve Blank, 29 July 2022

269 https://www.csiro.au/en/news/News-releases/2020/Speech-to-the-National-Press-Club

270 https://www.csiro.au/en/news/News-releases/2020/Speech-to-the-National-Press-Club

271   https://www.csiro.au/en/news/News-releases/2020/CSIRO-sets-sights-on-new-team-australia-missions-program

272   https://www.csiro.au/en/news/News-releases/2020/CSIRO-sets-sights-on-new-team-australia-missions-program

273   https://www.csiro.au/en/news/News-releases/2021/Fuelling-a-clean-and-bright-future

274   https://www.csiro.au/en/news/news-releases/2021/$150-million-missions-to-boost-australian-agriculture-and-food-sectors

275   https://www.csiro.au/en/news/news-releases/2021/$150-million-missions-to-boost-australian-agriculture-and-food-sectors

276   https://www.csiro.au/en/news/news-releases/2021/$150-million-missions-to-boost-australian-agriculture-and-food-sectors

277   https://www.csiro.au/en/news/News-releases/2022/CSIRO-on-a-mission-to-end-plastic-waste

278   https://www.csiro.au/en/news/News-releases/2022/CSIRO-on-a-mission-to-chart-Australias-low-emissions-future

279   Interview with Roy Green, 23 August 2022

280   Interview with David Thodey, 8 August 2022

Index